The ELECTRIC Image

Examining Basic TV Technology

M.A. Krupnick

Knowledge Industry Publications, Inc.
White Plains, NY

The Video Bookshelf

The Electric Image: Examining Basic TV Technology

Library of Congress Cataloging-in-Publication Data
Krupnick, Michael A.
The electric image: examining basic TV technology/Michael A. Krupnick.
p. cm. -- (The Video bookshelf)
ISBN 0-86729-296-2. -- ISBN 0-86729-282-2 (pbk.)
1. Television. I. Title. II. Series.
TK6630.K75 1990
621.388--dc20

Printed and bound in the United States of America

10 9 8 7 6 5 4 3 2 1

To Liz, the mother of my son
and
to Orson Max,
who will see wonders beyond my wildest dreams.

Contents

List of Figures....vii
Foreword....ix
Introduction....xi
List of Acronyms....xii

1 Perception....1
2 Scanning Theory....9
3 NTSC Color....25
4 Signal Switching and Mixing Theory....39
5 Magnetic Recording....55
6 Videotape Editing....75
7 Audio Sweetening....95
8 Digital Designs....105
9 System Mechanics and Troubleshooting....121

Glossary....133
Index....141
About the Author....145

List of Figures

Figure 1.1: Waveforms of AC Single-Cycle Sine Wave 4
Figure 1.2: The Cathode Ray Tube 5
Figure 1.3: Time as Geometry 6

Figure 2.1: Odd-Line Interlace (7-line scan) 13
Figure 2.2: Iconoscope Pickup Tube 15
Figure 2.3: Drive Waveform 16
Figure 2.4: Horizontal Sync/Drive Phase and Linear-rate Display 18
Figure 2.5: Vertical Blanking Interval, Sync Period (field1/field) 20
Figure 2.6: Interlace and Vertical Blanking 22

Figure 3.1: RF Modulation 27
Figure 3.2: RF Sidebands and Envelopes 31
Figure 3.3: Phase Delay by Wavelength 33
Figure 3.4: Reference Burst and Chroma 34
Figure 3.5: NTSC Color Encoder 35
Figure 3.6: Signal Timing and Picture Position 37

Figure 4.1: Functional Block: Switcher M/E Bus 41
Figure 4.2: Keys 44
Figure 4.3: Triple Re-entry Switcher 48
Figure 4.4: Switcher Sync/Video Processing 49
Figure 4.5: Matrix M/E Bus 51
Figure 4.6: GVG Model 300-3A Video Switcher 52

Figure 5.1: Electromagnetism 56
Figure 5.2: Magnetic Recording Head 58
Figure 5.3: Hysteresis 60
Figure 5.4: Gap Width/Signal Strength 61
Figure 5.5: Writing Speed 63
Figure 5.6: Quad Head and Threading Path 66
Figure 5.7: Quad Headwheel Servo 67
Figure 5.8: Quad Record 68
Figure 5.9: Quad Track Format 68
Figure 5.10: Quad Playback Signal System and Servos 70

Figure 6.1: Quad Splice 77
Figure 6.2: Quad Electronic Insert Edit 79
Figure 6.3: Time Code Phase 82
Figure 6.4: Edit Decision List 83

Figure 6.5: Distributed Processing Architecture 85
Figure 6.6: Helical Head Wraps 88
Figure 6.7: Helical Head Scanner/AST Detail 89
Figure 6.8: Ampex VPR-5 Portable Videotape Recorder 92

Figure 7.1: Beta Hi-Fi Format 101

Figure 8.1: Microprocessor Design 109
Figure 8.2: A/D Sampling 113

Figure 9.1: NTSC Scan Curve Distortion 125
Figure 9.2: User Data Time Code 128
Figure 9.3: Standard User Bits and SuperBits 130
Figure 9.4: Edit Decision Lists/Code Formats 131

Foreword

The document you are holding is the product of several years' work with a number of different schools and corporate entities, a few hundred students and a select group of broadcast professionals.

The schools involved with the evolution of this text are the University of Southern California Cinema/Television School and the Institute of Video Technology, Los Angeles. Both provided the incentive and moral support that produced the first spark of inspiration for this task.

The invaluable contribution of KTTV television, through it's willingness to open a broadcast engineering school, gave the project its first life-breath.

For many years, steadfast encouragement and personal involvement came in generous measure from KTTV veterans Walter Kalb, Lee Pardee, Robert Sudock and Charles Amman.

The Electric Image would never have made it to print without Liz Newstat. She was merciless as the proofreader, tireless as the editor, doggedly determined as the agent and ruthless as the salesperson. This book belongs to her as much as to me.

This instruction is unlike any previously published for television technical operations. It can be most informative when the reader recognizes and remembers that fact. *The Electric Image* is NOT an electrical engineering textbook, a production "how-to" guide or an equipment manual. We approach television theory from a physics standpoint, combined with a historical perspective of the reasons for engineering practices and conventions. This study will attempt to connect the reader more directly with the mechanics and evolution of the television signal by transcending electronics, instead of becoming obsessed with them.

You will notice that a curious mixture of topics is covered, for a "first level" study. Initially, some subjects within a chapter may appear too elementary, while others may seem too advanced. The fact is, to understand television theory you need to have a firm grasp on the subtle aspects that are rarely (if ever) covered in the usual basic studies. The traditional view is that certain topics are too difficult to coexist in a single lesson, even if they are intimately connected in theory. When these fundamental connections are overlooked, the student is left to fill the information vacuum, often with misconceptions.

The perception of television as physics can provide a number of very efficient detours around such traps in traditional instruction. This, of course, eases further study and enhances comprehension of more advanced topics (such as electronics in television) later on.

But, this broader learning comes at a price—it is not a predictable or familiar path. The reader is cautioned that the shortcuts are already built into the text. DO NOT SCAN, SKIP OR JUMP AHEAD! The chronology of the lessons is important and the program of study is totally unlike conventional texts. You cannot double-guess the path, so resist the temptation; it will build more quickly than you'd think from past experience. Keep your mind open and free of obstructions, and take it one page at a time. Let the book guide you through the side door and you will be rewarded with a view of the inside of your TV that is both stunning in its complexity and comprehensible in its simplicity.

M. A. Krupnick
1990

Introduction: The Forest For The Trees

Television engineering is a tough nut to crack. Ask any professional broadcaster; many will readily admit that they haven't cracked it even after decades in practice.

Education in technical operations, though offered by colleges, trade and tech schools, and a dizzying assortment of books, is for the most part, woefully inadequate. Most students get little more for their efforts than a headache.

After 20 years of radio and television operation and five years of teaching, I believe I've uncovered the biggest single obstacle to understanding television theory—electronics.

Not the study of it, but the assumption that electronics is the only environment in which TV theory can be taught. This leads almost always to a course in electronic theory. The study gets so hopelessly sidetracked that the main goal—understanding how television works is all but forgotten in the confusion, and learning is effectively derailed.

This book is about broadcast technical theory. It's about video, not electronics. You and I will stay on track by avoiding distractions wherever possible, including the monster-electronics. We will substitute simple physics for electronics, mechanical models for equations, and examine the history of today's broadcast TV system.

You will discover that television is amazingly simple when you look at it as a physical system. Videotape recording makes perfect sense when you can appreciate its evolution. Understanding this peculiar beast is like viewing a painting: you see it better when you step back a pace or two.

Let's take a long look at the electric image in a non-electrical way. We will end up knowing much more about one of man's most ambitious miracles—broadcast television.

List of Acronyms

The following is a list of acronyms used throughout this book.

CD	Central Dynamics
CE	Callaway Engineering
DEC	Digital Equipment Corp.
EECO	Electronic Editing Corporation
FCC	Federal Communications Commission
GVG	Grass Valley Group
ISC	Integrated Systems Corp.
NAB	National Association of Broadcasters
NTSC	National Television System Committee
PAL	Phase Alternating Line
SECAM	Sequential Colour a Memoire
SMPTE	Society of Motion Picture and Television Engineers
STL	Studio-Transmitter-Link

1 Perception

Humans perceive the world around them through their senses. The gathering and reproduction of information for television and video productions usually relies on the use of two of these senses: hearing and sight. In this chapter we will look at how the properties of light and sound are broken down electrically and reproduced as electronic signals that are transformed by our brains to be understood as sight and sound.

SOUND

Sound is generally described as a series of differential pressure waves traveling through a medium, usually air or water. These waves result from the vibration of an object, which transfers its motion into the surrounding matter. The pressure waves are reflected, partially absorbed, diffracted and otherwise influenced by objects and energy as they travel, until they reach our ears, where they cause similar vibrations in our eardrums. In the ear, these vibrations are changed into electrical energy for transmission to the brain via the auditory nerve.

Sound is measured in two dimensions: loudness (volume) and pitch, which is the height (depth) of tone.

LIGHT

Light is generally described as a member of the family of electromagnetic radiations—the broad spectrum of radiations between radio waves (both natural and man-made) on the low end and cosmic rays on the high end. Between these extremes lie such familiar forms of energy as heat (infrared light), visible light, ultraviolet light, X-rays and nuclear radiations. Electromagnetic radiations are often described as subatomic particles called photons. In quantum mechanics, photons are emitted by atoms which have absorbed external energy. Photons travel in groups that behave like waves. Because photons are units of pure energy, they can move through a vacuum (or space). Sound waves are pressure waves traveling through a volume of matter; they cannot travel through space. Light rays are bent, reflected and/or partially absorbed during their travels until they reach our eyes, where they are focused onto a photosensitive membrane called the retina. The optic nerve converts the light into electrical signals that are passed on to the brain as vision.

Light can also be measured in two dimensions: luminance (brightness) and chrominance (color). Chrominance is further specified by saturation and hue.

Saturation defines the purity or richness of a color and is expressed as a percentage of the total light energy. For example, if 20% red plus 80% white equals 100% pink, the color saturation of the pink would be the amount of red it contains, or 20%. White has a saturation value of zero.

Hue describes the tint of a color. It can be linked to the pitch measurement in sound—the red end of the spectrum is low in "pitch," and the blue-violet end is high.

ELECTRICITY: THE COMMON DENOMINATOR

Though light and sound are quite different in nature, they share one important attribute: they are both converted into electricity by our bodies before we can sense them.

Electricity, defined as electrons in motion, has the following properties:

- Voltage—the electromotive force (EMF) of travel; the amount of force moving the electrons. (Similar to the pressure of water in a pipe.)
- Current—the number of electrons passing a given point in a second. (Corresponds to the number of gallons of water in the pipe at a given time.)
- Resistance—the force of drag against the movement of the electrons. (The friction between the pipe's inner surface and the water droplets.)

Electrical Signals

Electricity displaying an organized configuration is called a signal. Such signals can be used to carry or communicate information. Consider, for example, the difference between the smoke of a campfire, which contains no organized message, and the same smoke, manipulated by an Indian brave who is sending smoke signals to his tribe. He may be sending a message that antelope are nearing the mountain, which has no relationship at all to the actual smoke or fire.

Electrical signals carry information by displaying regular shapes. These shapes are usually expressed as mathematical functions of normal Euclidean geometry, and are designed to model whatever action the electricity is performing. Simple waveshapes or waveforms include sawtooth triangular, rectangular and sine curved waves. (See Figure

1.1.) These waveforms are measured in two dimensions:

1. Wavelength—the length (in meters) of one cycle of the signal.
2. Amplitude—the height of the signal (in volts).

Wavelength has additional expressions of *frequency* and *period*, derived by performing some simple math. Frequency is the number of waves passing a given point in a second. Period is the inverse of frequency, or the amount of time it takes for one cycle of the wave to pass that same point.

The frequency and period of a wave are computed by using the known speed of light in a vacuum—186,000 miles per second. Electron particles, which are heavier than light photons, travel through a vacuum at nearly the same speed. When electrons travel through wires they are slowed slightly by collisions with the conductor's atoms.

It is important to note that even though all length measurements are easily convertible into their different expressions, none of these has any relationship to wave amplitude, which defines a signal's strength.

Applying Signals

The applicability of a signal (like any artificial model) is a direct result of how accurately the signal is constructed to serve as a symbol of some naturally occurring object or action.

Waveforms are representative of energy delivered over time. Wavelength is a measurement of the time frame of the delivery and amplitude is a measurement of the amount of energy delivered.

Consider the dynamics of the different shapes and how they translate to common everyday experience. Which basic wave shape best describes the gradual rise and fall of the sun? What shape is indicative of the way a lamp is switched on for an hour's reading, and turned off when that hour is over? Is there a shape that shows the pain when you stub your toe on the corner of a chair?

THE WAVEFORM: TIME AS GEOMETRY

The Oscillograph

A waveform can be tailored in shape to describe any function, simple or complex. The classic waveform plot, time (X-axis) at a right angle to strength (Y-axis), can be recorded with the oscillograph. Commonly seen in the doctor's office as the electrocardiograph, the oscillograph charts the simple rhythm of your heart. As part of a seismograph the oscillograph can also aid geologists by recording complicated earthquake data.

The oscillograph is really nothing more than a mechanical plotter—a roll of paper

Figure 1.1: Waveforms of AC Single-Cycle Sinewave

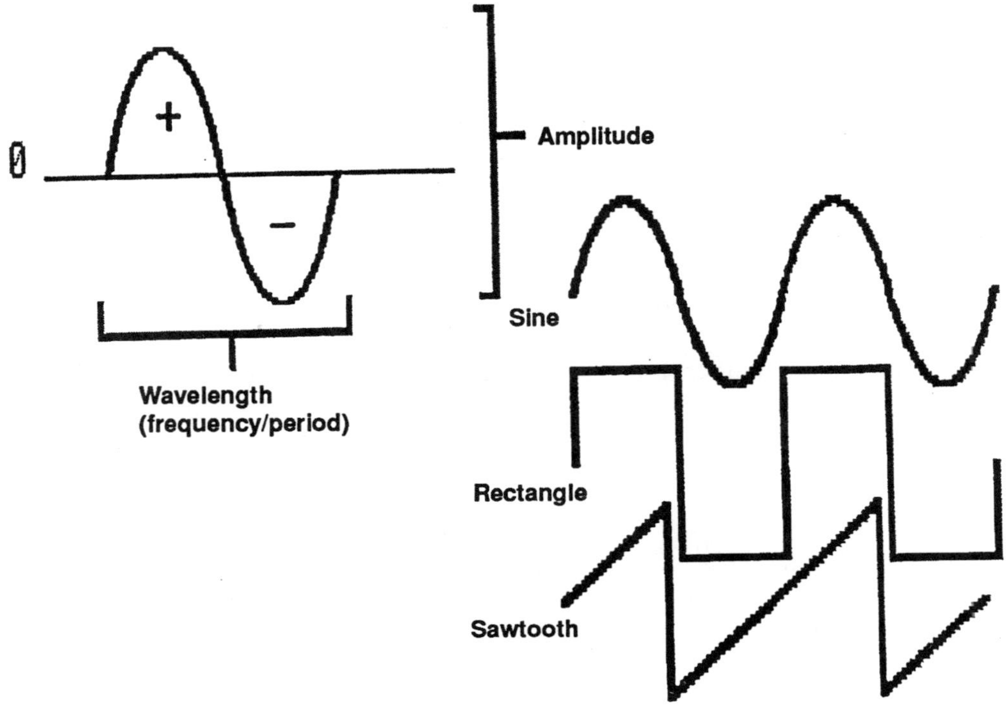

moving horizontally (the X-axis) under one or more pens swinging vertically (the Y-axis). The combined movements graphically depict the form of the wave. However, swinging pens have physical speed limits, and rolls of paper are finite in length.

The Cathode Ray Tube

Replacing the pen and paper with a vacuum tube, the Cathode Ray Tube (CRT), extends the useful range of measurement. The CRT is faster than the oscillograph because it uses electrons to draw the plot. This makes it possible to examine waveforms of complex shape and shorter lengths. The CRT was the forerunner of the television kinescope tube, or kine, as it is commonly called.

How The CRT Works

CRT operation is very simple. The CRT is an evacuated glass bulb with an electron source (cathode) at one end, and a target for the electrons at the other. (See Figure 1.2.)

The cathode emits a stream of electrons, each having a unit of negative electrical charge. The electrons are mutually repellent, and are attracted to any part of the tube having a less negative (positive) charge—the target end, the walls, the top and bottom, in short, anywhere but the cathode end.

To keep the electrons on the path to the target, there are a number of control grids placed along the way that are fed by separate sources of current. These grids surround the beam with a negative "wall" that keeps the electrons from escaping through the side-walls of the tube and accelerates them forward to the target. By carefully balancing and manipulating the grid currents, the beam can be precisely positioned to fall at a pinched point on the target.

The X-axis corresponds to the current directed at the right/left grids (horizontal drive); the Y-axis corresponds to the top/bottom grid current (vertical drive). When these currents are equal, the beam falls on the target at dead center.

Figure 1.2: The Cathode Ray Tube

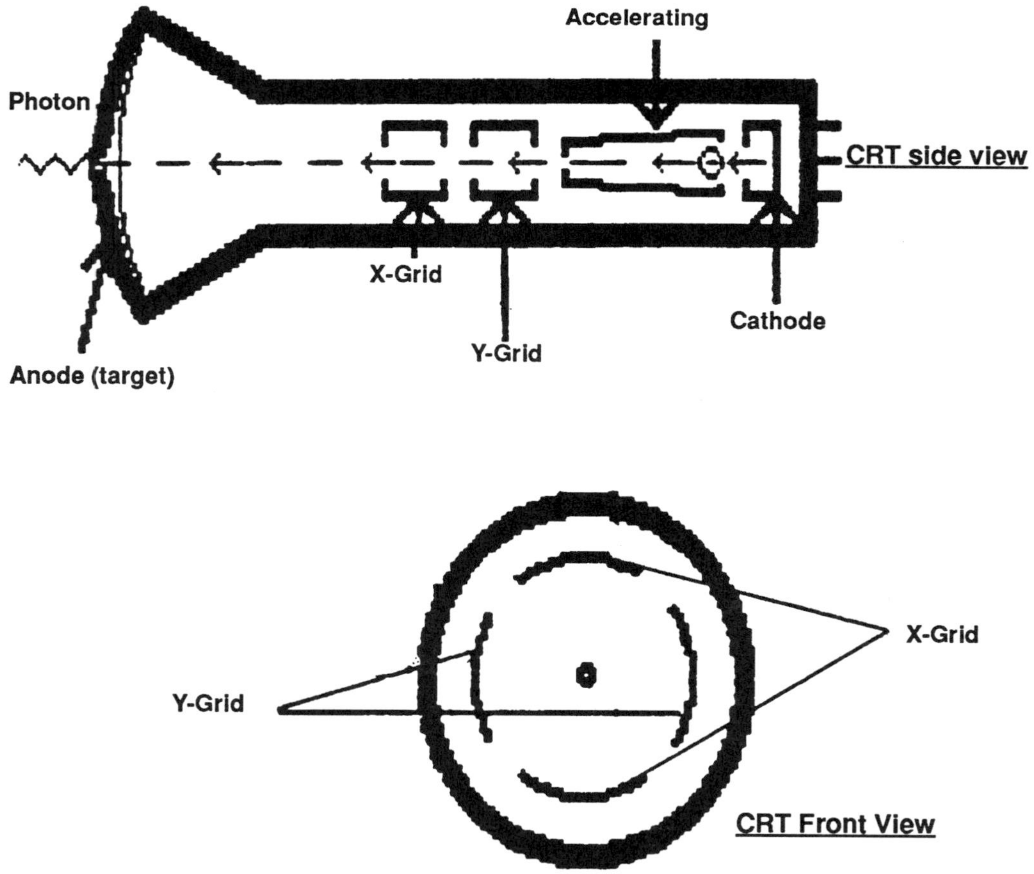

Figure 1.3: Time as Geometry

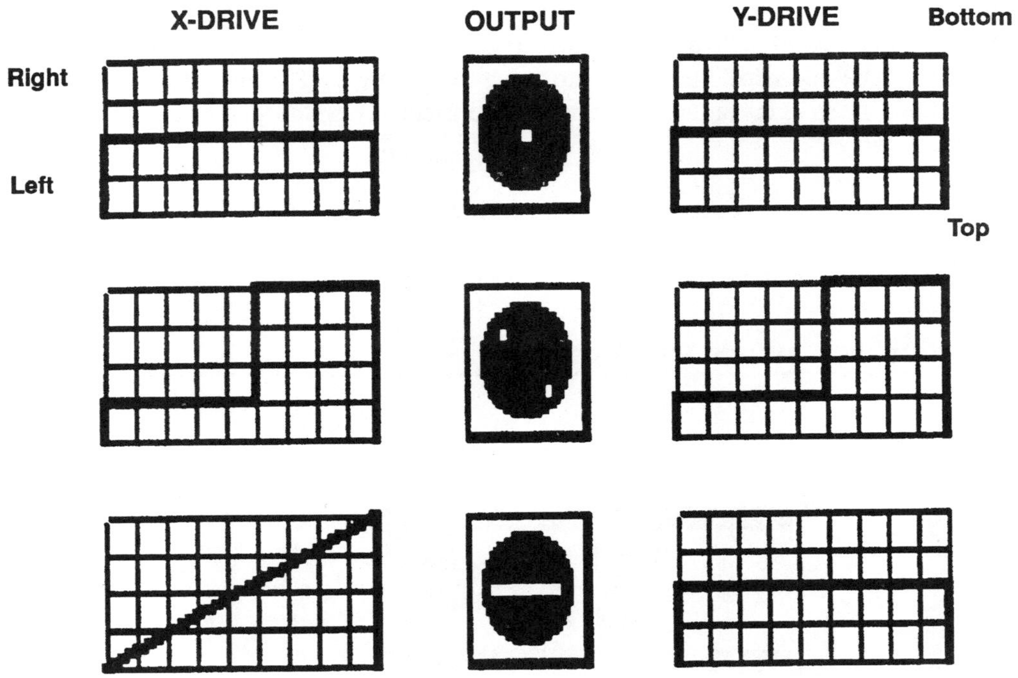

The target or anode end (most positive) of the tube is coated with phosphors. Phosphors are minerals that react to electron bombardment by releasing photons of visible light. The result is a bright spot (dot) where the target is struck by the electron beam. If the drives are offset vertically or horizontally, the beam and the spot land at a corresponding off-center position.

If one varies the value of the drives at uniform rates, the dot will move across the display face of the tube. This action is called sweeping the drives. The higher the frequency of the sweeping waveform, the faster the movement, until the dot no longer appears to be a dot, but a line. By using different combinations of drive waveforms and by varying the cathode current (which makes the dot brighter or dimmer), it is possible to draw a picture made of light—using electrons as the paintbrush. (See Figure 1.3.)

The patterns of light displayed on the face of the tube are the result of two types of signals: the positioning signals (X and Y) and the brightness signals (cathode current). Sending these signals by wire or radio to identical CRTs will communicate the patterns—in other words, transmit pictures.

CONCLUSION

The study we are conducting in this book is a chronicle of the problems encountered when man attempts to send information representing pictures through the air. The remarkable inventiveness with which the many obstacles were met, and the final goal realized, will be discussed in later chapters. Chapter 2: Scanning Theory, focuses on the development of television and the scanning standards set by the National Television System Committee (NTSC).

2 Scanning Theory

Television was invented early in this century as a mass medium for information exchange. The tools of the trade have increased and mutated from day one, but the methods of transmission and the structure of the signals have remained virtually unchanged. The television signal has been kept compatible over the decades as broadcasters added information to it without altering its basic structure. As a result, the system has a number of standard fixed specifications which are prohibited from becoming obsolete.

During the early days of television, although the quality of the images was carefully considered, quality was always secondary to the ease of installation of the entire broadcasting system. This was of particular importance considering the environment at the time. Radio transmissions were already firmly in place on assigned frequency bands worldwide, and telephone and telegraph networks were globally established. TV transmissions could not be allowed to interfere with them. The idea was to add a new communications channel without disrupting those channels already available.

Another important consideration was control of the monumental expense that was expected in implementing such a novel system. Television was, therefore, designed to deliver acceptable quality for minimal investment. Many of the limitations of current TV systems arise from these considerations.

THE SCANNING CRT

The original design for the scanning CRT display tube, also known as the kinescope (kine), came from two different scientists working independently: Philo Farnsworth and Vladmir Zworykin. Zworykin's design was of interest to young David Sarnoff of RCA, and the corporation worked to influence worldwide acceptance of his basic design. (RCA was the parent company of NBC.)

As the name implies, the design for the scanning CRT is merely an elaboration on the basic oscilloscope CRT.

The oscilloscope CRT, designed as a measuring instrument, displays only a single line. In the oscilloscope, the X-axis drive defines the time base for measurement and the Y-axis drive (the measuring input) plots the amplitude variations taking place as the signal is traced. Since the scope is really little more than an electronic ruler, the cathode current never varies. The dot is always illuminated, even when no signal is present on the Y-axis. The dot is usually swept underneath a calibrated measuring grid—the graticule—which is inscribed on a clear faceplate in front of the display.

In the kinescope, the objective is to display a picture, not a graph. So, the kine CRT displays a stack of individual horizontal lines. During the trace of each line, the scanning dot is varied in brightness to represent the illumination of that portion of the picture. This varying cathode current is the *video signal.* The measured amplitude of this signal is the *luminance* or brightness of the picture. The standard measurement is .714 volt peak-to-peak for 100% white on the screen, and .053 volt for black. Note that screen black is not zero volts (no signal). When the CRT is turned on, there is always some illumination of the screen. Since the kine stacks lines, it must have a constant signal, even during dark periods. We all know what a TV set looks like when no signal is fed to it.

This line stacking requires rapid sweeps of the dot forming the horizontal lines. Each successive line must be stacked uniformly below the previous sweep as the scanning progresses. To accomplish this, a slower vertical sweeping action is used to produce a neat and uniform stack. The result is straight (linear) traces which vary only in brightness (unlike those of the oscilloscope). Each line is the same length and starts at the same distance from the left edge of the display. The vertical spacing of the lines is uniform and consistent.

In the kine, the horizontal (X-axis) drive defines the length of all lines, thus the width of the picture. The vertical (Y-axis) drive defines the rate at which each line is added below the previously drawn line, therefore defining the height of the picture. (Hereafter, we shall refer to the X-axis grid as the horizontal deflection coil, and the Y-axis grid as the vertical deflection coil. Likewise, the drive signals for these coils will be referred to as horizontal and vertical, in keeping with the convention of geometrically describing the electrical picture elements.)

CREATING STANDARDS

In the late 1940s, the National Television Systems Committee (NTSC) convened to formulate standards for television broadcast systems to be operated on this continent. Representatives from Canada, Mexico and the United States were present. They considered the basic requirements which had to be satisfied to ensure that all TV receivers operating in North America would have access to all broadcasts originating within that continent. The NTSC standards developed at that time are basically unaltered today.

It is noteworthy that the other current television systems—PAL and SECAM, widely used in Europe—are no more than modifications of the basic NTSC structure.

THE NTSC SYSTEM

Because communications channels for other media had already been established, the new television channels had to operate within broadcast bands, using "air" not occupied by police, fire, commercial radio, military or telephone traffic. TV was to operate on the 60 hertz (Hz) power standard prevalent throughout the continent, and had to present a picture and sound that would be acceptable to a public familiar with 16mm motion pictures.

Examining the last point first, 16mm film has a *resolution* (the amount of detail perceptible in an image) of approximately 450 lines/square centimeter. This means that a test pattern of lines shot on 16mm film is readable as separate lines by a normal viewer until 450 of them cover one centimeter of the negative. After that, the lines fuzz together, due to the graininess of the emulsion.

During a line scan in the CRT, a test pattern of vertical lines is represented by a video signal that alternates between white and black. The more lines, the more rapidly the signal must change, causing the frequency of the video signal to become a limiting factor.

Because television had to be kept within already-defined bandwidths—6 megahertz (MHz) per channel—the maximum horizontal resolution attainable at the time was about 350 lines. This resolution would have to be equal in both vertical and horizontal directions.

The original *aspect ratio* (height to width of the displayed picture) of 16mm film is 1.33 to 1, or three units high by four units wide. The television display copied this ratio.

Motion pictures exploit the image retention properties of the eye by showing a rapid succession of still pictures to the viewer. The rate is 24 film frames per second. Television, based on the 60-Hz power standard, shows 30 frames each second.

The picture is displayed as a *raster*, or group, of lines forming a rectangle resting on its long side.

Each television frame is composed of 525 horizontal lines, scanned from left-to-right and stacked from top-to-bottom. This grouping takes into account overscanned lines (lines not displayed as picture) and delivers a vertical resolution capability that is so nearly equal to the horizontal resolution as to be symmetrical. This allows details in the image to be equally sharp in both dimensions.

The combined vertical (slow) and horizontal (rapid) sweeping of the dot makes each line trace slightly diagonal, higher on the left and lower on the right side of the display.

The scanning and spacing of these lines defines the exact location of each single *picture element*, or *pixel*, the illuminated point on the screen made by the dot. The pixel is the smallest unit of the picture. Though the dot is theoretically a circle, the rapid scanning action tends to smear the dot slightly, making it an ellipse in practice. This horizontal elongation also limits the resolution of the system.

During early viewing tests, an annoying flicker was apparent in the picture. This was due to the amount of time between frames when no image was being scanned on the screen. To reduce the amount of "black" time, the obvious solution was to increase the frame rate. Bandwidth limitation requirements prevented this, however, as increased frame rates would cause the channel to overlap into protected bands of assigned broadcasts.

Odd-Line Interlace

To compensate for this problem, the NTSC realized that the frame rate could appear to be increased by scanning the same number of lines twice as fast, halving the "black" time between frames and reducing the flicker to levels imperceptible to most viewers. Each frame was, therefore, scanned twice—all the odd-numbered lines first, then all the even lines. The frame was divided into two equal fields, odd (1) and even (2). Dividing 525 evenly meant that each field would contain 262 and one half lines.

This method of scanning is called odd-line interlace. It is characterized by an odd total of lines scanned an even number of times (525/2). All television systems broadcasting today use odd-line interlace. The reason we see no systems with an even number of lines is that the two fields are identified by differences in the positions of the resulting half-lines.

Unfortunately, this method of reducing image flicker while preserving bandwidth has produced a number of ambiguities and problems which, like the standardization the NTSC gave us, stay with us to this day. We will be discussing some of these ambiguities in later chapters.

In Figure 2.1, we can examine the basic scanning structure of an NTSC frame in a simplified form. Although only seven lines are considered, the physics are the same, and hold true for all odd-line broadcast interlace systems. You can see that the vertical retrace following the field one scan divides line 7 in two, with the first half treated as odd, and the second half as even because it occurs in field two. These two halves are separated by the screen's height, and are treated as different fields, even though they are parts of a continuous signal waveform. That is, although line 7 appears to be halved because of its screen position, it is not split in time. Line 7 is half odd and half even; and its position in the picture starts at the bottom and ends at the top. The same condition applies to NTSC line 525, or PAL line 625, or any last line in any odd-line interlaced scanning scheme.

The benefit of this trickery is a 30 frame picture with the apparent smoothness of a 60 frame motion rate. Interlaced scanning solves the problem of flicker using the same number of frames—a clever way of improving the picture without affecting the transmission channel. Today, many people complain that the NTSC television system is not as good as it could be. Interlaced scanning is one aspect of the system that is subject to much criticism.

Figure 2.1: Odd-Line Interlace (7-line scan)

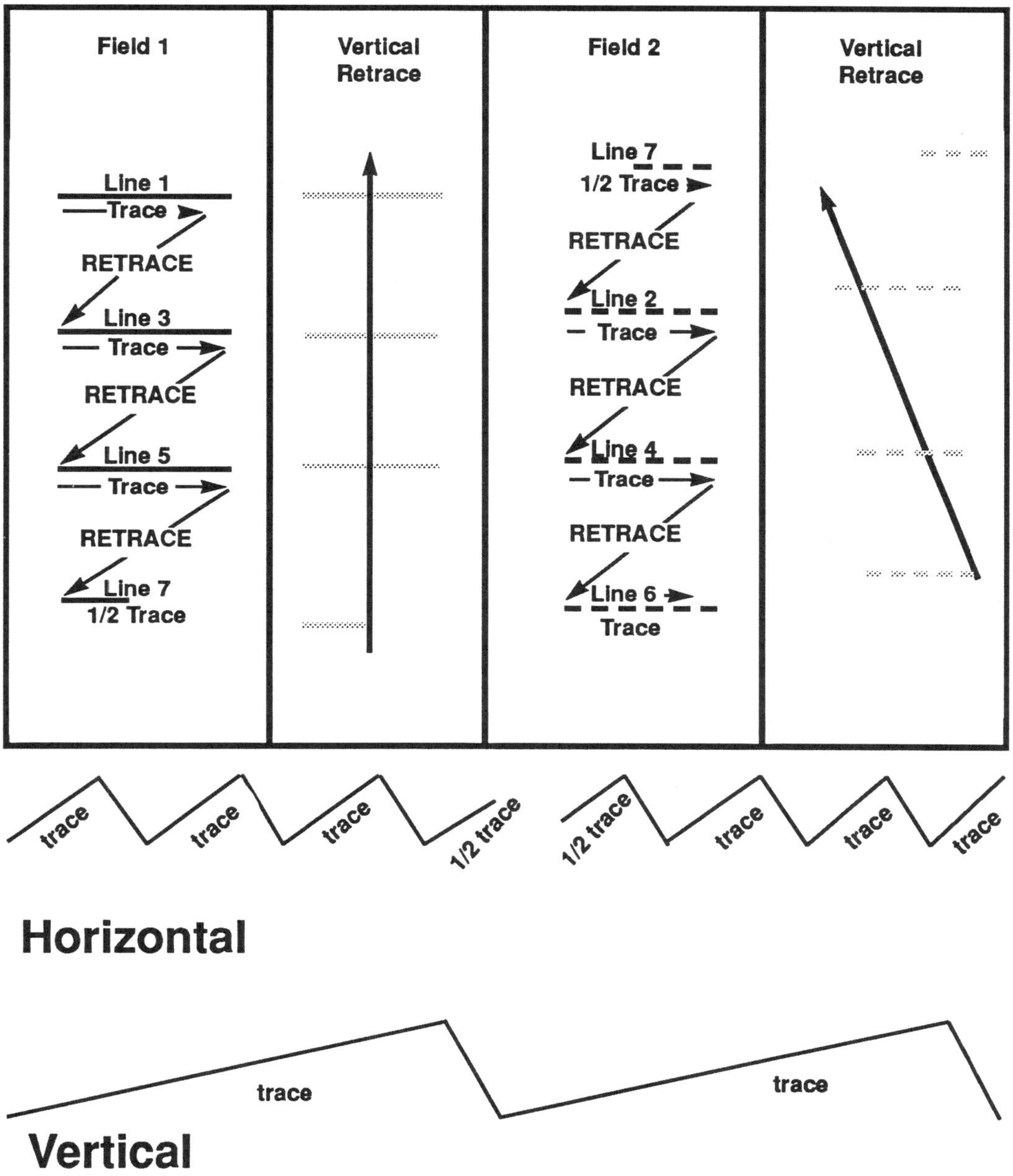

Many technicians prefer the advantages of improved technology such as computers or high-definition television (HDTV) systems that produce pictures with better resolution and quality. While this is certainly true, the NTSC system and its cousins have been broadcasting pictures since the 1940s—the newer systems are limited, so far, to closed circuit (wired) applications. Different schemes that were developed to reconcile the several high-definition TV (HDTV) designs have not been standardized. As a result, HDTV is still years away from being broadcast over the air.

It should be mentioned that the NTSC is probably the only multinational authority that has met its original goal without major international turmoil, and its resolutions have survived intact throughout decades.

The following list outlines a few simple rules for a standard NTSC picture:

1. Each frame contains 525 horizontal lines.
2. Each frame begins and ends with a full line.
3. Each frame is divided into two equal fields.

LINEAR SCANNING AND SYNCHRONIZATION

Linear scanning was developed as an attempt to transmit a group of signals serially—that is, one after the other, to produce a vertically stacked group of lines.

Obviously, the transmission would have to be extremely rapid if it were to attain acceptable results. Also the timing of each signal in the raster would have to be precise.

On the screen, each line would have to start at the same left edge as all the others, and be exactly the same in length. Failure to achieve uniformity would result in pictures with jagged edges and sheared images. To accomplish this, each line had to be stacked a uniform two lines' width (accounting for interlace) below the previous line to prevent vertical distortions of the images.

During testing, it quickly became apparent that each line's signal arrival time at the receiver determined the geometric screen position the line would occupy.

There had to be a predictable way of converting time into geometry for the system to display a meaningful picture—the signal being the communication of that geometry between camera and receiver, and all other receivers displaying the picture.

First, the luminance characteristics of a scene appearing before the camera had to be encoded into a train of time-dependent signals representing a scanned picture. This was accomplished by scanning the scene in the same way it was intended to be displayed. This action is called uniform linear scanning, and must be precisely repeated at the receiving end for the picture to be recreated.

Scanning a Scene

In the television camera, an image is presented through conventional optics to a pickup tube which is basically a kinescope in reverse (see Figure 2.2). Where light falls on the face of the tube, a photosensitive coating reacts with the photons and emits an electron for each photon. The electron falls on a charged grid behind the faceplate, where it neutralizes the spot on which it lands, creating a "charge hole." The grid is continuously scanned by an electron "gun." A precise set of sawtooth-shaped drive waveforms are used to control the scanning.

When an electron from the gun falls on the target grid's charge hole, the faceplate electron (called the *photoemissive electron*) exits on the grid output line as video: a signal representing a bright spot in the scene. The target electron exits the tube only when the scanning beam impinges on it. Since the gun is not "looking" at the whole scene at once, but scanning it as a series of vertically-stacked horizontal lines, the video output of the tube is the time-segmented serial signal we want.

Figure 2.2: Iconoscope Pickup Tube

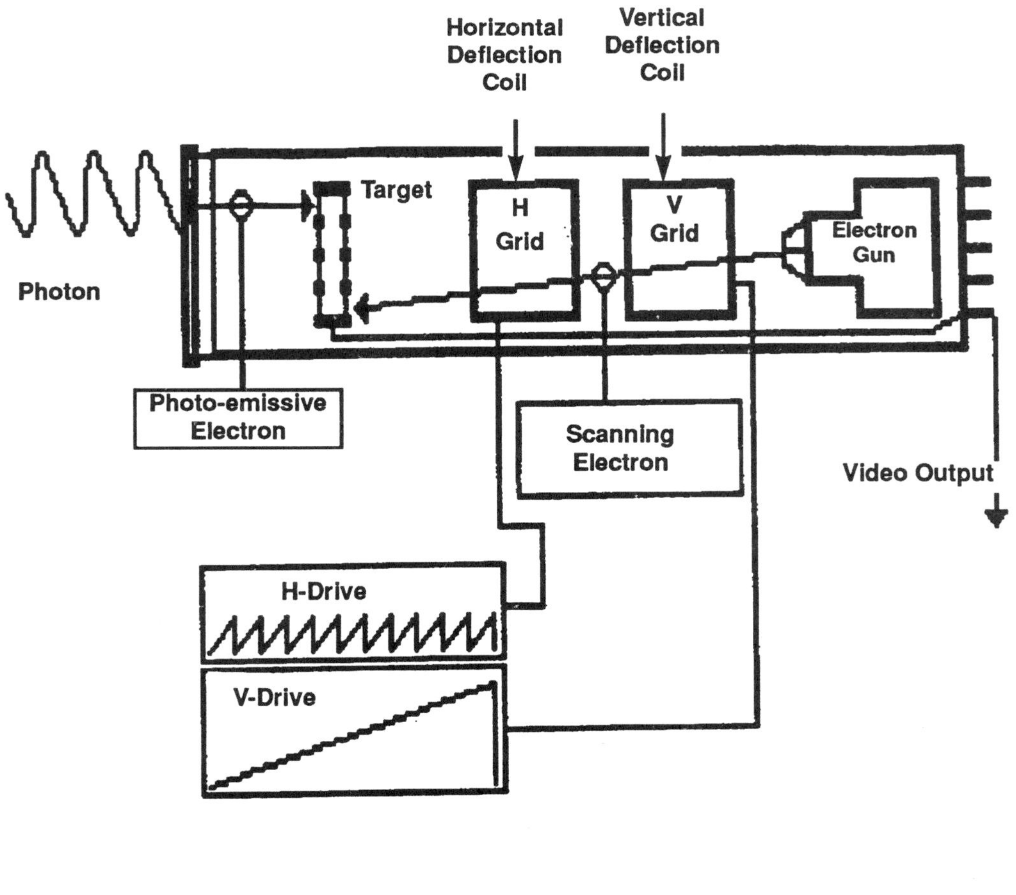

The timing of the scanning gun in the pickup tube determines which electron is picked off the target grid at what position. This is the key element of encoding the picture luminance values into sequentially transmitted signals, timed in such a way that converts quickly and accurately into a spot of measured light positioned on a map with one million possible locations.

If this process cannot be exactly repeated—in reverse—the signals will not be decoded at the kine (or receiver) to become pixels at proper geometric screen locations. Remember that the video signal itself is merely a varying voltage representing the brightness of the dot at a given instant. Without a uniform linear scanning reference to use as an index of screen position, the dot is merely a dot, not a pixel.

The uniform reference is provided by the sawtooth sweep pulses, or drives (see Figure 2.3).

The sawtooth driving pulses fed to the deflection coils represent the time base for the display of the picture.

The rising slope of the horizontal drive waveform represents the left-to-right active scanning period. The trace represents the visible scan of the horizontal line. From the peak of the signal to the trailing edge, the beam is turned off, thereby blanking the line as it retraces and reverses the scan from right-to-left. It then returns to the starting point at the left edge.

Figure 2.3: Drive Waveform

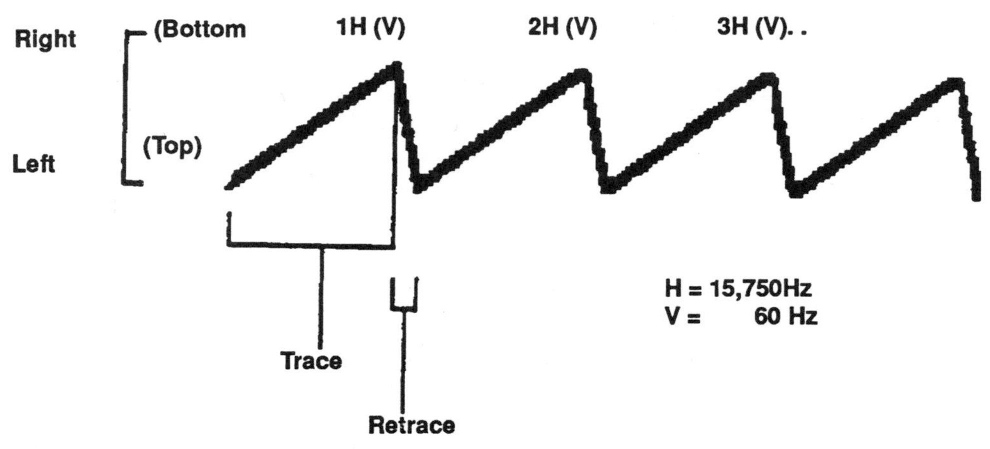

This sawtooth is the horizontal drive signal, and represents the horizontal time base. The frequency of the waveform is simple to compute: we draw 525 lines for each frame, and there are 30 frames every second. Five hundred twenty-five lines multiplied by 30 frames equals 15,750 Hz.

Meanwhile, another sawtooth signal is driving the beam down the face of the tube and defining the vertical time base for the spacing of each line in the stack. This scan is slower and runs from the top to the bottom of the screen. The vertical drive signal controls this scan, and differs only in that it is fed to the vertical deflection coil and, of course, is a different frequency: 60 Hz for the 60 fields per second.

Since the function of these signals is the same, the waveforms themselves are also nearly identical, only the timing and destination differ.

The amount of electrons exiting the tube through the video line defines the brightness of each pixel. The time of the electrons' exit compared to these drive pulses defines the *position* of the pixel.

If the camera and receiver are driven by the same scanning pulses, the picture will be displayed free of geometric distortions.

Synchronization

Inside the studio, extensive wiring and signal routing systems provide one set of drive signals to all cameras, monitors and other production equipment so the pictures are scanned in perfect time coincidence.

At home, of course, this luxury doesn't exist. The TV receiver must make a picture using a signal literally plucked out of the air.

Horizontal Sync

In order to conserve the limited bandwidth available to a television broadcast channel, the drive signals cannot be transmitted. Although each receiver contains its own drive amplifiers furnishing very stable signals for scanning, there is no guarantee that each set will lock on to the broadcast at the right time to display the picture as it was scanned by the studio camera.

Therefore, each broadcast line is accompanied by a pulse timed to occur during the *horizontal blanking period*, the time during which the line is in retrace. This signal is added to the video waveform each time a drive pulse is generated in the studio. It marks the line's synchronizing point for the receiver and is called the *horizontal synchronizing pulse* (see Figure 2.4). The pulse is a rectangular wave whose leading edge coincides with the peak in the studio's sawtooth. Like all sync pulses, it starts at the "clamping line" (0%) and drops to a standard level of -40% relative to video. Since it occurs during blanking and is blacker than picture black by almost 50%, it cannot be seen. The duration of the pulse represents the time needed to retrace back to the left of the home screen. There is a short period of blanking after retrace, represented by the horizontal back porch, extending from the trailing edge of sync to the leading edge of pedestal or setup at +7.5% above clamping. This period is a safety margin, included to ensure retrace is finished before active scanning signals representing the next picture line are transmitted.

By using horizontal synchronization, the TV set at home is brought into time coincidence with the line (or horizontal) timing at the studio without the use of a separate drive channel. The TV set shifts the timing of its horizontal drive to match the timing of the incoming horizontal sync signals.

Vertical Sync

When the studio finishes a field scan and is in vertical retrace, a vertical synchroniz-

Figure 2.4: Horizontal Sync/Drive Phase and Linear-rate Display

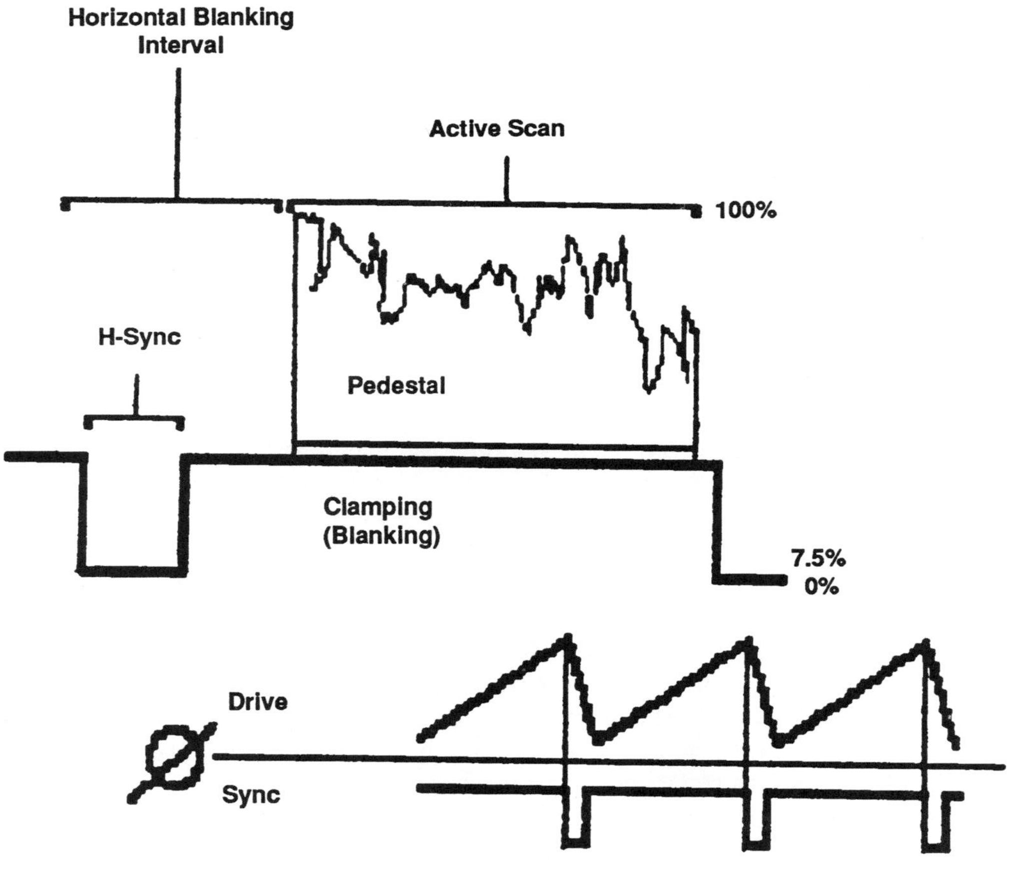

ing pulse is transmitted to provide the same timing stability for the vertical sawtooth of the receiver. Because these sync pulses occur in blanking, they are not visible. (Every line is accompanied by a horizontal sync pulse, and each *field* is accompanied by a vertical sync pulse.)

The trace/retrace sequence for the horizontal scanning process is straightforward, making the synchronizing requirement quite simple. The vertical scanning process is more complex because it includes the requirement for spacing each horizontal line two scans downward to allow for interlace. The interlacing action also results in half-lines—lines half the horizontal length of a regular line.

Since the NTSC standard demands a full line at the beginning and at the end of each frame, these half-lines are placed at the field division point occurring at the end of field one and at the beginning of field two, at the halfway point of the frame. In this way, the system is able to discriminate between *fields* and *frames*.

This is infinitely more complex than the film analogue of merely pulling down the next frame with a pin or claw mechanism, though the end result appears the same to the viewer.

At the end of the odd field, 262 and one half lines have been traced, numbered 1,3,5,7,9...525. Note that line 525 is not a full line, and is not the last line in the frame. At this time, the scanning dot is at the bottom/center of the raster, and the scanning beam is blanked for a vertical retrace (see Figure 2.5).

The signals sent to initiate the blanking are called the pre-equalizing pulses. There are six pulses sent in a three-line period. These pulses are therefore said to have a 2H (two times horizontal) frequency. The 2H rate of the pulses acts to keep the interlace circuits synchronized during the time that no active scanning information is being transmitted.

Following the pre-equalizing pulses is the *vertical synchronizing pulse*, also called the *serrated pulse*. It resembles the train of pulses preceding it. It is 3H in length, and is divided into six serrations, like the pre-equalizing pulses. The vertical sync pulse is differentiated from the pre-equalizing pulses by the greater amount of time these serrations are negative (at the -40% sync level). The function of vertical sync is drive synchronization (as with horizontal sync) but, in this case, of the vertical retrace action. During the transmission of vertical sync, the TV receiver should be retracing to the top of the raster. The next scanning field is the even field, so the retrace should end at the top/center.

After vertical sync, another train of pulses identical to the pre-equalizing pulses is transmitted. These are the post-equalizing pulses, of value in the old tube-type RF transmitters for maintaining interlace. For some time now they have been unnecessary, but are still transmitted.

The nine line period just described is known as the vertical *synchronizing period*, the

Figure 2.5: Vertical Blanking Interval, Sync Period (field 1/field)

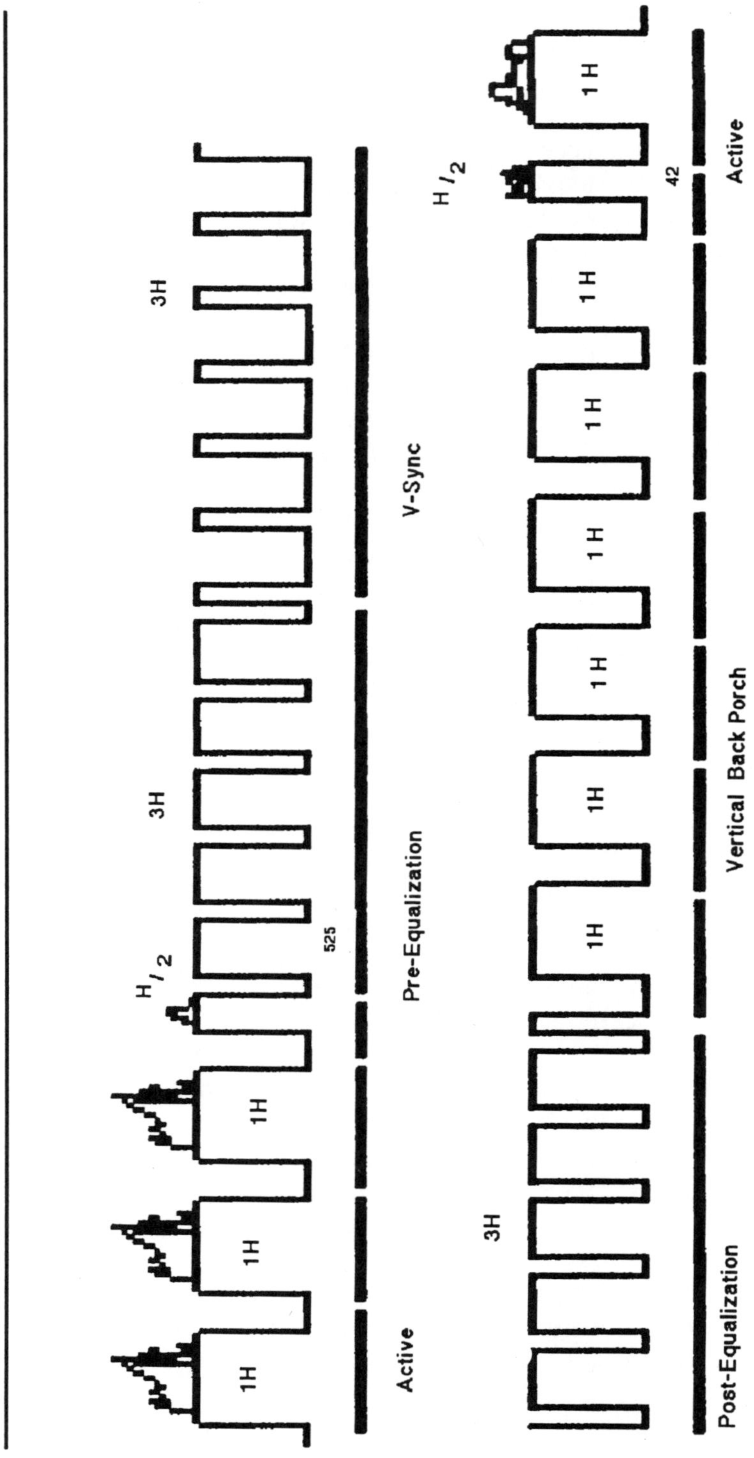

time during vertical blanking in which all vertical synchronizing information is transmitted. By the end of this period, the downward scan of field two should have begun.

At this time, a *vertical back porch* is transmitted for the rest of the blanking interval, consisting of clamping and horizontal sync to re-establish proper DC picture voltages and timing references for upcoming active lines. Specifications for the total length of vertical blanking are a minimum of 20 lines and a maximum of 21 lines per field.

The vertical back porch is often used to transmit or record a wide variety of signals with the picture. Vertical Interval Reference Signals (VIRS) are pulses indicating luminance and chrominance levels for automatic correction equipment. Vertical Interval Test Signals (VITS) are used to quality check transmission channels. Vertical Interval Time Code (VITC) is employed instead of, or in addition to, audio-frequency time code for automated editing and production systems. Closed captions (CC) are inserted in the back porch to display subtitles for hearing-impaired viewers. Because the back porch is blanked at home, information contained within it can be transmitted and received without interrupting or displacing the primary broadcast information.

The first active line scanned in field two is line 42 or 44, depending on the accuracy of the transmission system. This line is active only one-half the duration of a normal line. The active scan is where the half-lines are joined in the NTSC 525/2 interlace. Line 525 is half active, half blank. Sync and blanking are counted as full lines, because they have to be consistent from field to field. So, line 2 is a full line in wavelength, although it is not a picture line. It is part of the vertical sync period. To mate properly with line 525, line 42 is half blank, half active. The signals are continuous and unbroken; it is the active display which has "halves."

To further confuse the unwary, standard engineering publications from SMPTE (The Society of Motion Picture and Television Engineers), the FCC (Federal Communications Commission) and even manufacturers generally do not maintain strict adherence to the odd/even line count. When the vertical blanking interval is described, these papers often count the lines in the period from 1 to 21, regardless of the actual sequential scanning position. It is common to find references to, for instance, "line 19, field 2," or "line 18, both fields."

At the end of field two, (line 524) the frame is complete, and another vertical retrace is initiated preparatory to a new frame starting with field one.

This retrace begins at the bottom/right of the raster. The same series of vertical pulses and back porch are transmitted again, but this time the retrace ends at the top/left of the raster, half a line earlier than the previous field. The difference between the blanking intervals for the respective fields is the length of the active lines that precede and follow them. It is this difference (full or half-line) that allows the system to discriminate between a field blanking and a frame blanking. (See Figure 2.6.)

Figure 2.6: Interlace and Vertical Blanking

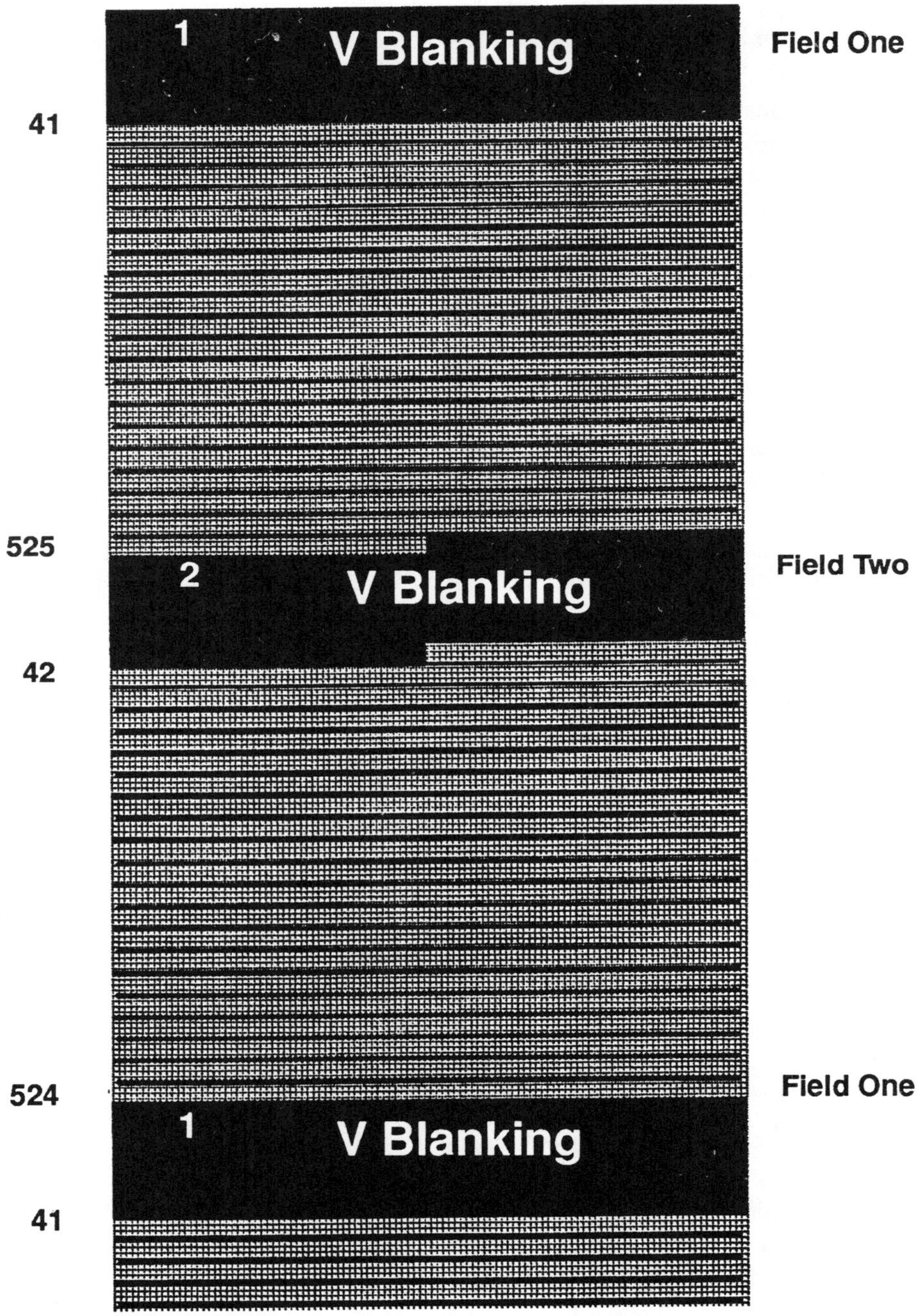

CONCLUSION

It is obvious that in television engineering practice, timing is everything. The period for one line at 15.75 kilohertz (kHz) horizontal rate is 63.5 microseconds. During that time, about 83% of the line is active picture information. The other 17% is synchronizing information and is not displayed. During the active scan period of 52.7 microseconds, the scanning beam forming the dot which is drawing the picture line is able to change from peak white to deep black, or any of a number of intermediate gray values, at least 350 times. Very tiny timing errors in signals can turn into gross picture distortions. The remarkable aspect of all this is that this mechanism was invented in the 1930s, and implemented on a wide basis long before the advent of the "electronic" age, or the "computer" age, and so constitutes the first high-speed reading/writing medium of man.

Furthermore, the basic design is unchanged today from the original prototype. Now, however, due to great advances in television networking and distribution totally independent of the original broadcast criteria, we are looking critically at the limitations of the NTSC system and contemplating for the first time radical changes in the mechanism.

Using the computer, the laser and other recent technological innovations, it seems certain that the face of television will be altered dramatically over the next decade. But it is just as probable that the changes will be additions on the sound fundamental structure we have grown to appreciate as the the classical NTSC system—outdated, yes, but as elegant and exciting an invention as man has ever wrought.

3 NTSC Color

It can be seen from the previous chapters that television was designed as the simplest possible mechanism for encoding pictures into signals suitable for radio transmission. The signals used to drive the transmitter are of two basic types:

1. Active Signals—those signals used to drive the video circuits of the receivers. The amplitude variations are translated into brightness variations of the scanning dot as it "draws" the scene on a pixel-by-pixel basis along each line.
2. Synchronizing Signals—those signals used to control the motion of the scanning dot, the uniform and linear sweep of which defines a time base. The signals are transmitted sequentially and converted into precise geometric locations for the dot on the CRT's display screen.

The system that was standardized by the NTSC was monochrome (black-and-white). Color wouldn't arrive until later. Before we move into an evaluation of NTSC colorimetry, it is best to review the basic aspects of RF (radio frequency) transmission systems. The nature of broadcast transmissions has an important impact on the understanding of television color.

RADIO FREQUENCY TRANSMISSION

When you call to someone at a distance, you raise your voice to a shout. If the distance is too great, the atmosphere absorbs the differential pressure waves created by your shouting before they can reach your target. Even very loud noises in the audio frequency band, such as sirens, are absorbed rather rapidly in the air. Sound is the motion of molecules, and is not an efficient way of transmitting signals through the atmosphere.

Electromagnetic radiation, on the other hand, travels rather efficiently through air. Radio is a member of the family of electromagnetic radiations, fields of energy that travel where sound cannot. The fields move around rather than through molecules of air, causing the atmosphere to be transparent to radio waves. Radio waves are very responsive to other electromagnetic fields, however. If air molecules have an electric charge caused by a bolt of lighting, for example, they become less transparent and disturb radio broadcasts. An electromagnetic field accompanies any electric current. When electrons at high frequencies and power are trapped in a length of wire, their energy produces a very strong electromagnetic field radiating outward in all directions. The wire becomes an antenna, emitting radio waves. The radio spectrum is high enough in frequency to be inaudible, and low enough to be invisible. But the wave can be detected by a wire of identical length (resonance), and is capable of carrying information great distances.

Radio broadcasts exploit these properties by translating audio-frequency (AF) information into radio-frequency (RF) information. This is accomplished by using an RF signal at a constant frequency or amplitude as a carrier for the weaker audio signal. The two signals are added together, the resulting sum being the *RF envelope* which is amplified and then transmitted. The addition is accomplished by a circuit called the *modulator.* This modulated carrier signal is used to drive the transmitter, an amplifier of high power. The power is connected to the antenna where it produces very high currents which alternate direction rapidly (thousands to billions of times a second). The atoms of the antenna are shock excited by the changing current. In this condition, the antenna emits a corresponding magnetic field which passes the shock on to electrons in the air, thereby charging the molecules. These molecules produce identical fields, which charge neighboring molecules, creating more fields. The radio wave moves outward at the speed of light, going great distances before it loses much power.

The receiving antenna is a wire that encounters the radio wave. The electrons in it are charged as the field passes, and a tiny current is induced in it. The antenna current is amplified and processed by the radio receiver.

At the receiver, a circuit called the demodulator is tuned to the carrier's frequency and subtracts the radio energy from the envelope, leaving the audio program signal. The audio signal is then amplified and appears at the speaker as sound.

Amplitude Modulation

The classical method for RF modulation is *amplitude modulation* (AM). The program signal is added to a carrier. The result is that the amplitude of the carrier signal changes as the program's amplitude changes (see Figure 3.1). The frequency of the carrier is constant. The envelope of an AM transmission is an RF carrier of fixed frequency with changes of its amplitude at AF rates. The receiver passes this signal through a tuned circuit resonant to the carrier frequency. In essence, the circuit filters out the higher radio frequency (RF) component of the signal, but passes lower audio frequencies (AF) through. *Demodulation* is the subtraction of the carrier from the envelope, the opposite of modulation.

Figure 3.1: RF Modulation

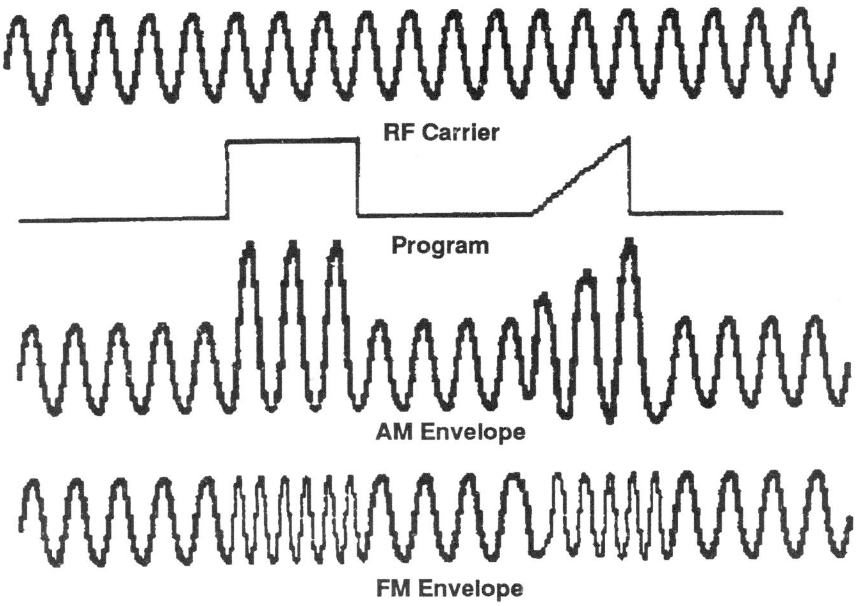

This type of demodulator, called a detector, is sensitive to changes in amplitude, and outputs the program—in radio as an AF signal. Amplitude Modulation means change in height. In AM, the amplitude of the RF envelope carries the program information.

AM is used in television to broadcast the picture signal (at video frequencies extending to 4.2 MHz) on the visual carrier.

AM is sensitive to amplitude changes, so anything that introduces amplitude changes in the transmission path generates interference on an AM channel.

Frequency Modulation

Frequency Modulation (FM) differs from AM in that the carrier, when modulated, changes in frequency as the program's amplitude changes. An FM envelope is a carrier of fixed amplitude whose frequency changes at program rates. The demodulator used for FM is the discriminator, which ignores amplitude changes (thus being insensitive to AM interference), but senses the amount of frequency change in the carrier. In FM, the frequency of the carrier contains the program information. Television sound is broadcast using FM. A television transmission has always been an AM/FM simulcast.

NTSC Regulation

RF modulation of signals produces a set of sidebands, high energy fields of alternating magnetic polarity (N/S) and electric charge (±).

These modulation products have frequencies that are multiples of the combined envelope (RF carrier + program = envelope). For example, if a 10 Hz program signal is modulating a 100 Hz carrier, the envelope consists of upper sidebands at 110 Hz, 220 Hz, 440 Hz and lower sidebands at 90 Hz, 45 Hz, 22.5 Hz, etc. These sidebands are the signals that the receiver must decode to present the program signal.

In the case of NTSC television, it was realized early on that the simultaneous transmission of aural (sound) and visual (picture) carriers would produce a broader spectrum of sidebands than ever previously transmitted on a broadcast band. Strict adherence to limitations in the bandwidth of transmitted signals was imperative to prevent overlap of sidebands and the resultant inter-channel interference.

Though color TV was only a theory when the NTSC first met, the committee was aware of the fact that future transmissions would have to be compatible with the original signals. So the NTSC added another criterion to its basic standards, "the system must be compatible and reverse-compatible regarding color." Neither part of the system—transmitter or receiver—would make the other obsolete when updated to newer designs, especially the big jump to color.

This requirement has been honored through the years remarkably well. Nobody had to junk a TV set because a particular station switched to color, and today's receivers make old kinescope recordings look better than the day they were made.

COLOR BASICS

Anyone who has marvelled at the magical beauty of a rainbow is familiar with the reality, if not the theory, of additive color. The color of light comes from the frequency of the photon emissions that make up the light. The emissions can come from nuclear reactions, like those in the stars; kinetic reactions like the glow of a lamp's filament; or chemical reactions like those that light up the belly of a firefly. Light interacts with objects and forces it meets along its path. A mirror placed in the path of a beam of light will reflect it, changing the direction in which it travels. Light rays are bent during their travels through refractive materials, like water or lenses. A certain proportion of light energy is absorbed by virtually every object in its path. The sunlight falling on a peacock, for example, is white—it contains every color of the spectrum. The light reflected back from the bird's plumage, however, is blue-green, because the feathers absorb the red light.

Note that there is a difference between *incident colors* emitted directly from a source, and *reflected colors* that bounce off an object. Reflected colors survive absorption during the "bounce." These are subtractive colors, the colors of pigments, and the most familiar to us. In this color scheme, red + yellow + blue = black. Crayons, inks and paints are subtractive in their colors. Colors coming from objects which shine with their own light (like those from the surface of the sun) are additive. The phosphors of a CRT emit their own light, making additive colors. Mixing additive colors results in white.

NTSC Primary and Secondary Colors

The NTSC color system has three primary colors, that is, colors that cannot be duplicated by the combination of any others: red (R), green (G) and blue (B). Mixing these three colors together in equal parts produces white. Black, of course, is the absence of light, and so, absence of color.

There are three NTSC secondary colors, those colors which result from the addition of two primaries: magenta (R+B), yellow (R+G) and cyan (G+B).

A color has two basic properties: hue and saturation. A color's hue is its relative position in the spectrum compared to the primaries. Its saturation is the amount of pure color present that is unadulterated with white. For example, a hue of cyan indicates an absence of red and the presence of equal parts blue and green. A pure cyan would have a 100% saturation (50% each: blue and green). If the color were pink, it would have a red hue, but a low saturation, since pink is red + white. Keep in mind the difference between chrominance values and luminance values: luminance is the brightness of picture, and chrominance is the color content. For instance, a picture of a whitewashed wall may have high luminance, but no chrominance at all, since white is not considered a color.

Continuing to think along this line, it could be said that white, though the sum of all primaries, can be regarded as the absence of color. White and black are luminance values only, and have no chrominance. There are shades of gray between black and white; they share the same attribute—no color, just luminance.

NTSC studies with test subjects indicated that human vision is most sensitive to greens (59% of white), least sensitive to blues (11% of white), and moderately sensitive to red (30% of white). Another important discovery was that human vision doesn't rely much on color to resolve detail in a scene. Though cyan and orange colors are somewhat resolvable, generally the detail in a picture is sensed by luminance (Y in NTSC terms) values, and the colors are "filled in," just like a coloring book.

NTSC color starts with these premises, and attempts, again eloquently, to make the execution of the system as simple as possible.

THE COLOR CAMERA

The color camera is basically three monochrome cameras in one chassis. The different pickup tubes are presented with the scene through optical filters that conform to the RGB/Y ratio of 30-59-11. When these values come from the tubes, the pixel is colorless—gray, white or black. Whenever one or more of the tubes is not outputting a signal in the RGB/Y ratio, the pixel has a color. The hue and saturation of that color are directly proportional to the departure value from Y.

Although the theory of the color camera is simple, the technical realization is far from it. Aside from the intricacies of dealing with the physics of photoconductive coatings and the inconsistencies of vacuum tubes, there is the painstaking attention to the geometry of the light path to the different tubes (which must be as close to identical as possible in all dimensions). Add to this the need for just as much absolute symmetry in the electrical operation of the camera, and it becomes clear that keeping three optical/electrical channels in perfect balance is no mean feat.

But, this is successful only in breaking up picture chrominance, or color value, into its component RGB parts. It is only a starting step toward color display of a picture—it doesn't begin to deal with RF transmission. And a weakness is present that effects both display and transmission: there is still no distinction between chrominance and luminance. For instance, how would the camera "know" the high red gain appearing in the R-tube when it "sees" an apple is really the color red and not just 30% of white? The same high red output would occur in the R-channel if the camera were shooting a white card.

Moreover, NTSC standards prohibit the transmission of separate R, G and B channels because of bandwidth excursions. Finally, to make things even more difficult, we must transmit the color information without affecting the luminance information to adhere to the criterion of color/monochrome compatibility. In other words, the apple cannot appear too black on a monochrome set receiving the color broadcast.

MONOCHROME TRANSMISSION CHARACTERISTICS

FCC rules dictate that each television transmission channel occupy a frequency band no more than 6 MHz in width, including both visual and aural carriers. Since the audio portion of a broadcast generates a relatively narrow set of sidebands (20 kHz maximum), we will not be concerned with it here.

The video signals, when modulating an RF carrier, generate most of the sideband information. As would be expected, the sideband energy is distributed at multiples of carrier + video frequencies; specifically, at harmonics of the frequency (carrier + line). The line frequency is 15.75 kHz. If examined on a spectrum analyzer, the RF information would appear as it does in Figure 3.2: the unmodulated carrier in the center of a symmetrical set of sidebands corresponding to energies of c+H, 2(c+H), 3(c+H), etc., on one side, and c-H, 2(c-H), 3(c-H), etc. on the other. Between these "clumps" of upper (additive) and lower (subtractive) sidebands, there is no energy at all. All of the luminance information is concentrated in the sidebands, and it is this information that is received and interpreted at the home TV set as a monochrome picture.

NTSC ENCODED COLOR

The modification of the monochrome system to color standards meant that the sin-

Figure 3.2: RF Sidebands and Envelopes

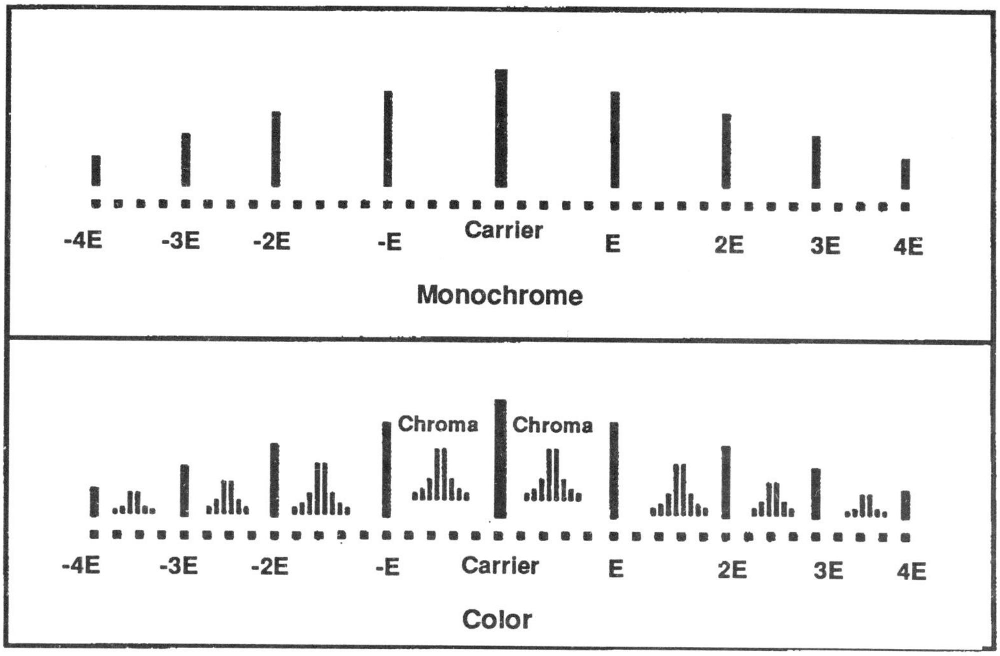

gle channel in the camera and the kinescope was to be replaced with a three-channel system: luminance would be treated as a ratio of red, green and blue channels of 30/59/11. The kinescope and camera would use three guns instead of one.

Though the addition of two more electrical/optical circuits in these components is more elaborate, it is not very complicated. Unfortunately, the signal was not to be transmitted directly as RGB, but merely translated to RGB at the beginning and end of the process. This is a complex matter, involving some form of information compression during the transmission to meet bandwidth limits, and re-expansion to RGB and Y after transmission.

The key to the puzzle is that any ratio of RGB (three channels of information) not conforming to 30 + 59 + 11 (Y) can be defined in two values of color difference information: hue and saturation.

The addition of color requires another carrier within the broadcast RF envelope. Because of bandwidth limitations, the modulation couldn't be FM; because of compatibility requirements, the color envelope could not interfere with the luminance sidebands already present in the Y portion of the RF envelope.

If the color carrier frequency is chosen as an odd multiple of one-half line (H) frequency, the sidebands for modulated color RF fall neatly in the null spaces in the Y +

carrier envelope, and will not interfere with the other modulation products of the transmission (refer to Figure 3.2). The result is two separate channels of picture information: luminance (Y) and chrominance (color-only information), all within the 4.2 MHz bandspread allowed for visual RF.

A monochrome receiver ignores the chroma sidebands; and if a monochrome picture is being transmitted without color carrier (hereafter referred to as subcarrier), the color receiver has no color difference signal to interpret, and "sees" the broadcast as Y-only.

The frequency of 3.58 MHz (rounded off) was chosen to accomplish these objectives. This frequency is perilously close to the upper limit of the envelope, and must not be modulated heavily under any circumstances. But how could this carrier be modulated to simultaneously represent two values specific to color signals?

PHASE MODULATION

The modulation of the subcarrier under the most extreme color conditions would have to be minimal, but at the same time, subtle differences in color shading would need to be translated accurately and faithfully.

The challenge was to create two detectable differences in a new carrier without expanding the envelope. The frequency and amplitude of the subcarrier could not change grossly under the influence of the modulating color difference signal.

The solution was *phase modulation* (PM). Instead of altering frequency as is done in normal FM, the subcarrier is delayed a fraction of its wavelength to represent a hue change. These small increments of delay can be expressed as time values, but they would become complex fractions, difficult to handle. If the delay is expressed as a part of a wave cycle, corresponding to the degrees in a circle (See Figure 3.3), then 360 different hues can be easily specified as degrees of rotation. More subtle hues can be expressed within this structure as minutes and seconds of arc.

In fact, the range of color selection available in NTSC television exceeds that possible in film. This form of measurement is called *polar mapping* or *vector analysis* and is useful in specifying small delay values. The use of PM instead of FM limits the bandwidth of broadcast chroma hue information with no sacrifice in the clarity of hue definition.

The subcarrier is amplitude modulated to represent saturation changes. The subcarrier envelope is effectively limited in its amplitude by offsetting the modulation index (the zero-modulation state of the carrier) to place all negative cycles of the signal below the threshold of luminance values during active scanning, and below the clamping line during blanking intervals. In other words, the chroma signal arises from the instantaneous value of luminance—it is *always* a 3.58 MHz sine wave centered on the luminance

Figure 3.3: Phase Delay by Wavelength Fraction as Expressed by Degrees of Rotation

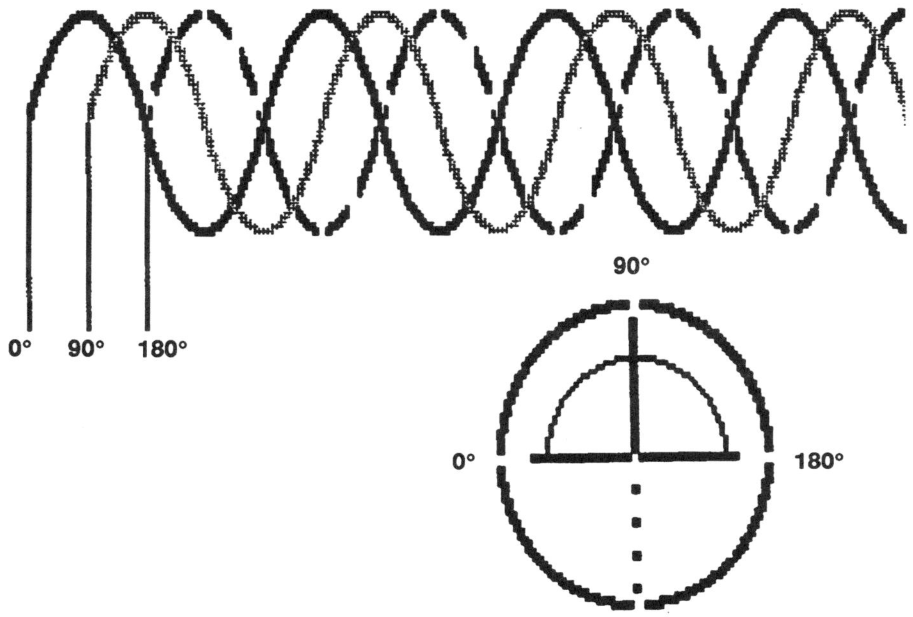

line. This way, the chroma levels are prevented from ever exceeding luminance levels. Chroma sidebands cannot extend over the Y sidebands in the transmission.

In order to provide a reference signal to compare against the modulated signal, *color sync* consisting of 9 to 11 cycles of unmodulated subcarrier is inserted in the horizontal blanking interval of each line. This is *color burst,* and occurs just after the end of the horizontal sync pulse (see Figure 3.4).

Color burst is used to signal the presence of chroma (modulated subcarrier) on the line, and is therefore included in every color broadcast. The reference level of 40% centered on clamping is used to encode saturation information. The absence of color burst signals a color-killer in receivers to turn off the chroma processing, preventing addition of noise to monochrome transmissions.

THE NTSC COLOR ENCODER

Now we have a coding scheme that is responsive enough to deal with subtle color values, yet broad enough to portray extreme color gradients without exceeding the confines of the defined RF bandwidth.

The camera and kine, however, due to the conversion from a single-gun system to

Figure 3.4: Reference Burst and Chroma (Magnified H-Rate Display)

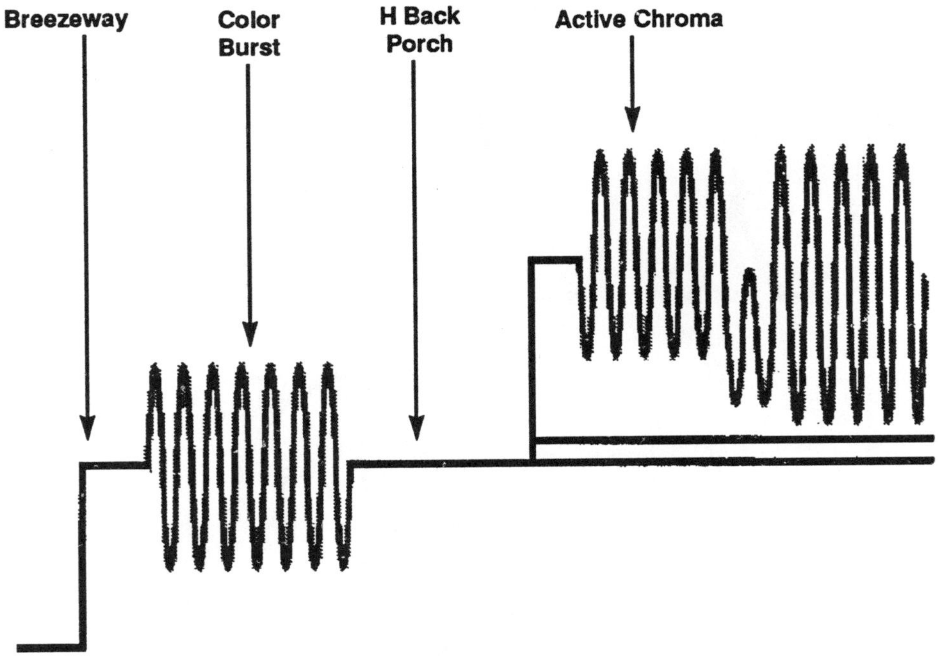

the three-gun RGB primary configuration necessary for color display, are now unable to do what they did formerly—present a Y signal as luminance-only information, that is, without chroma; a true monochrome. This is because we must now integrate R, G and B signals to arrive at Y. This must be done outside of the camera, using the RGB signals.

The encoder accomplishes this by passing the camera's RGB output through a passive network which adds the channels in the 30 + 59 + 11 proportions (see Figure 3.5). The network is the matrix of the encoder, and when Y values appear, the matrix passes the signal forward as integrated luminance (Y). If the sum is not equal to Y, the matrix integrates the RGB values into a color difference signal.

The color difference signal is fed to one side of a modulator whose other input is the subcarrier. The encoder phase modulates and amplitude modulates the subcarrier, using the color difference signal as the program source. The output is therefore modulated subcarrier (chroma). The phase of the output represents the hue and the amplitude represents saturation.

If a Y signal emerges from the matrix it triggers an inverter where the subcarrier is added to the inverted subcarrier, that is, subcarrier 180° out of phase. The resulting sum is zero: no subcarrier, no chroma.

The color information was finally truly separate from the luminance information

Figure 3.5: NTSC Color Encoder

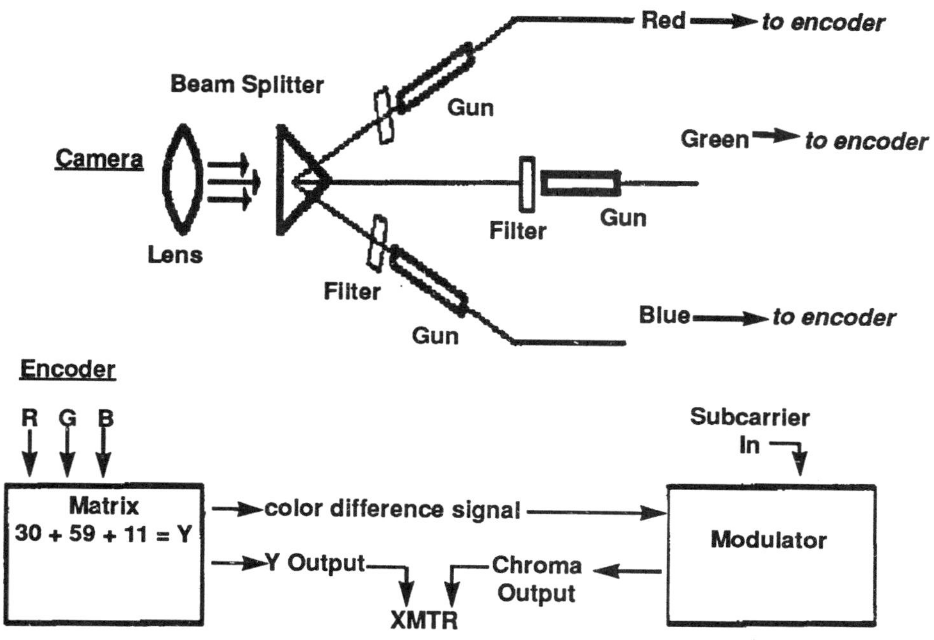

anywhere along a line, yet fully integrated into the broadcast signal transmissions.

Upon testing this complex scheme for broadcast-encoding color in a mono-compatible fashion, the NTSC and FCC discovered that one annoying flaw remained: the new subcarrier had a tendency to interfere slightly with the aural carrier in high luminance picture transmissions, producing a nasty buzz in the sound which varied directly with picture brightness.

To eliminate this, the line (H) rate was shifted slightly further away from the 3.58 MHz subcarrier frequency, from 15.75 kHz to 15.734 kHz. A shift in H rate, of course, inevitably shifts the vertical rate: from 60 Hz to 59.94 Hz for the field rate; and from 30 Hz to 29.97 Hz for the frame rate.

These small "corrections" in basic scanning rates were well within the lock-in range of even the primitive AFC (automatic frequency control) circuits that were used to drive receivers in the early days of television. Pictures exhibited no bad side effects and the sound interference disappeared. The modification was therefore adopted.

Much later, this "inconsequential" adjustment, like interlace, would return to haunt engineers. But in 1953, the formidable problem of mono-to-color transition in NTSC television had been solved—with brilliance.

A number of clever aspects were built into the new color system. One is that the subcarrier itself is suppressed. It is not transmitted as a continuous-wave RF signal, like the other carriers. The only time an unmodulated subcarrier is present in a broadcast is on the back porch of horizontal blanking, and then only for 9 to 11 cycles. Bandwidth is conserved because nearly no envelope space is taken up by a stationary carrier.

The choice of frequency for the subcarrier (odd multiple of 1/2 line) has the added side benefit of reversing the phase of the subcarrier on each line. Any spurious chroma noise on adjacent lines, or the same line in the next frame, cancels (0 phase + 180 phase = 0 signal), limiting the possibilities of noise in the signal becoming noise in the picture.

THE COLOR TIME BASE

As is already evident, the time-dependent nature of the various components of the television signal is used to define the geometric positions of the picture elements.

The easiest way to examine and measure these signals is with high-speed oscilloscopes (waveform monitors and vectorscopes). These instruments yield accurate displays of the period of each signal (see Chapter 1), and show how various parts of the complex waveforms are related in time. While there are volumes of specifications for these relationships, they make rather dull reading. For example, horizontal blanking is described as no less than 10.58 microseconds (µs), and no more than 11.4 µs in length, consisting of a front porch of 1.54 µs plus or minus 0.05 µs, followed by the leading edge of sync, with a fall time maximum of 115 nanoseconds (ns), and extending for no more than 4.71 µs—and the specification goes on for a page and a half.

In practice, what is just as accurate and far less cumbersome is to compare a signal to a known standard NTSC reference signal. By overlaying the two or switching between them on the scope, you can easily see a graphic display of any inconsistencies (see Figure 3.6). Standard timing yields a normal standard picture display. Any errors are expressed as departures from standard reference timing.

The widest time base is clamping level, which occurs as a "flag" during active scanning, and as the starting point of all signals during both active and synchronizing periods. Clamping is generally regarded as a fixed level DC (O Hz) reference for all other signals. The remaining complex waveforms are all AC type.

The field rate signals at 59.94 Hz define the vertical time base. A vertical signal arriving late at the receiver displaces the picture vertically—it rolls down the screen. Early vertical time base errors roll the picture up.

The horizontal rate signals at 15.58 kHz define horizontal position; errors displace the picture horizontally.

Color sync at 3.58 MHz is the highest frequency signal in the video spectrum:

Figure 3.6: Signal Timing and Picture Position

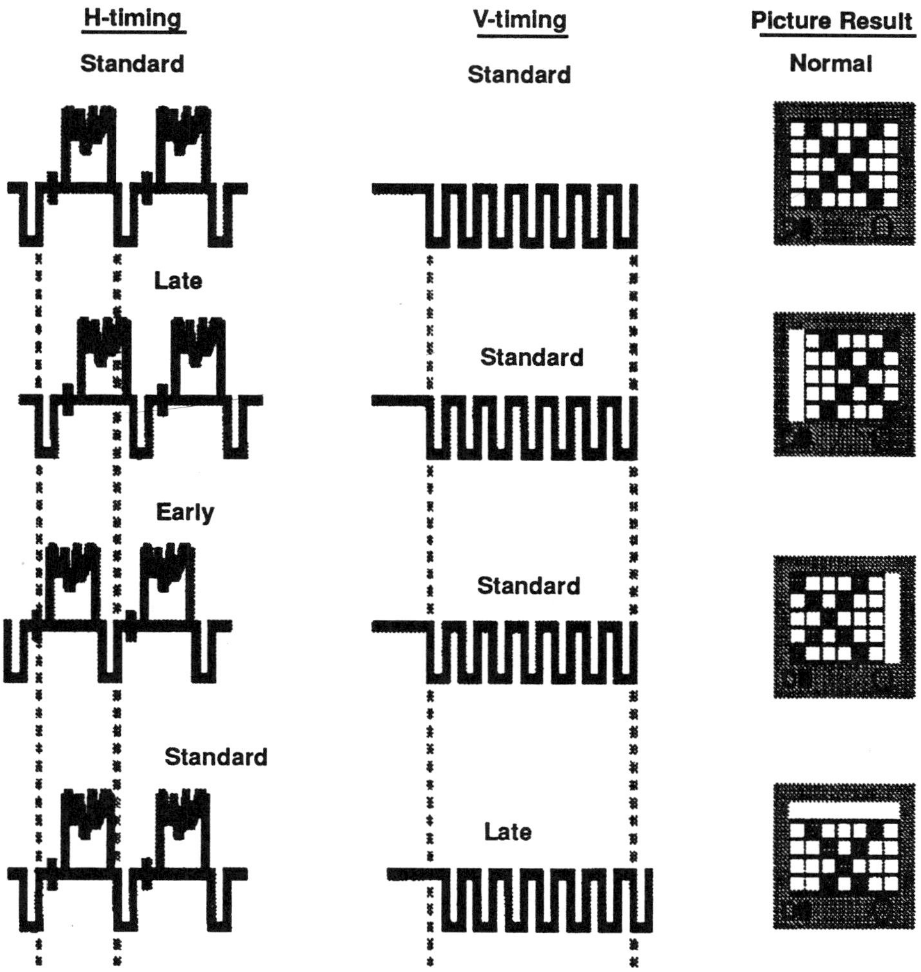

errors in the phase of subcarrier result in shifts of hues in the picture; errors in amplitude of the subcarrier result in saturation distortions. The high frequency of subcarrier makes very small timing errors (280 nanoseconds is one complete cycle, or 360° of phase rotation) visible in the picture.

The need for tight timing control of a color television signal is 1,000 times more exacting than for monochrome. This type of engineering precision is exhibited each moment a picture arrives in acceptable condition at the home TV set. The set is performing in excess of 60,000 operations every second. It is amazing that this precision is maintained so reliably and unnoticeably to the home viewer. It should be clear by now that there are more reasons for the picture to break up than to remain stable. It is a tribute to those who developed and participate in television technology that we do not fall down more than occasionally while skating on such thin ice.

4 Signal Switching and Mixing Theory

Before there was television, there was the motion picture. The visual "language" of motion pictures in the 1940s was quite highly evolved. The public, which had grown up watching movies, was the same audience that would be watching television. The quality of production that this audience was accustomed to would have to be carried over into television to make it a successful medium. It was not so simple as it sounds.

Most of the visual effects that are added to film to enhance storytelling and suggest abstract ideas are added during the post-production process, after the shooting has been completed. Television, on the other hand, was designed as a real-time medium—the broadcast reaches the audience at the instant it is shot.

Cinematic devices such as close-ups, medium two shoots, wide-angle perspective, dissolving of viewpoint to another place or time and special effects have always been physically necessary, since a camera seldom holds enough stock to shoot a full-length picture. Despite that, keeping the attention of an audience with one angle for as long as 90 minutes would be difficult at best. Film visual effects usually require weeks of editing.

Making the live television picture as exciting and entertaining as a movie without weeks of post-production was the challenge. The solution was to use several cameras that were integrated by the use of a video production switcher.

SIGNAL TIMING

The vertical blanking interval is the TV frame division point. If the switch from one camera to another occurs during vertical blanking, the on-screen effect is the same as a film splice—a clean cut. If the switch occurs at any point outside blanking, it will be visible on the screen. (Refer to Chapter 2 for a discussion of vertical blanking.)

Of course, the two cameras that are switching must *both* be in veritcal blanking during the switch. If they are not, timing difference between the scans will produce positional errors between the frames that can disturb the picture and possibly break it up.

Synchronizing Cameras

The best way to ensure identical scan timing is to drive all cameras with the same scanning signals. Usually, a television station or studio has a single master synchronizing pulse generator. This *sync generator* is a crystal-stabilized oscillator that produces a *master clock pulse* timing signal. The frequency of the master pulse, which is several times higher than the normal color subcarrier, is constantly monitored and corrected to maintain a consistent rate and waveform. The clock signal is used to generate a wide variety of pulses that are especially suitable as TV sync and reference signals. The timing and frequency of these final signals are derived by dividing the master pulse to create new waveforms.

The sync generator outputs horizontal (H) drive, vertical (V) drive, various blanking signals, burst, subcarrier, H-sync, V-sync and equalization, composite blanking (sync + burst + blanking)—in short, virtually any signal used in television equipment for timing purposes. The controlled stability of the master signal is passed on to all these signals as they appear at the generator's output.

In order to distribute these signals throughout a television system, the sync generator outputs are connected to a *pulse distribution system*, which amplifies each output and provides multiple feeds for additional equipment.

Timing and Cable Length

The pulse distribution system gets the signals to each of the cameras. Just as important, however, is the time at which the signals arrive. Electrons speeding through wires at nearly 186,000 miles per second would seem to be arriving instantaneously. In fact, the wire does resist electrical current flow, and has a measurable delaying effect on signals passing through it. The more cable a signal must traverse, the more the delay. Because we are dealing with very precise timing values, serious picture disturbances may result from the tiniest errors. For example, the right edge of the picture is less than 64 millionths of a second away from the left edge. If two switched cameras have cable length differences of as little as 10 feet, the cut between them may fail completely.

Originally, timing problems were handled by making all cables the same length. Since nothing can be accelerated faster than the speed of light, signals that might have arrived early from one camera had to be delayed to match the arrival time of later signals. This meant that every camera had to have the same amount of cable as the camera with the longest cable run. While this made the cable makers ecstatic, the cost of cabling an installation was dismaying to the station owner.

Fortunately, delay can be added less expensively by passing signals through resistance/capacitance (RC) circuits, which store electrons for a short, controllable amount of time before passing them on. These circuits and others, such as the "bucket brigade" analogue types and newer digital versions, are all classified as delay lines—those circuits that change only the time a signal is output.

SWITCHER CONSTRUCTION

While the video switcher is the most elaborate looking piece of equipment in the television control room, it is really one of the simpler devices used in television—at least on the level of theory. It functions in only two ways (switching and mixing), no matter how complex the setup or the effect appears.

As the name implies, the switcher "switches" video sources. It is the video equivalent of an audio mixing console, or the preamplifier/tuner in your home audio system. The switcher's basic simplicity of function is reflected in it's modular construction (see Figure 4.1). Though there are many buttons, they are grouped into three distinct processing stages, which may act upon signals singly or in combination. They are:

- Crosspoint switching,
- Summing or mixing,
- Keying and/or effects.

Figure 4.1: Functional Block:Switcher M/E Bus

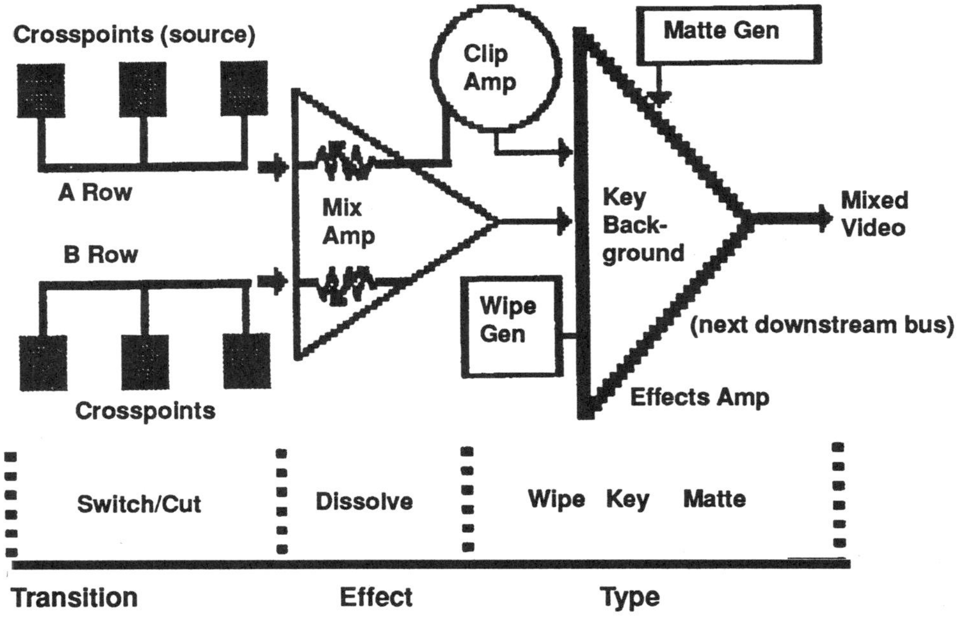

The video signals are presented to switching inputs, called *crosspoints*. All cuts take place in the crosspoint switching module of the switcher.

The summing or mixing amplifier is where two or more signals are added together, as in a dissolve or fade.

If a third signal is to be added, as in a title key, another section will come onto the processing chain, called the *switching amplifier*, or *effects amplifier*. This amplifier generates effects by using extremely fast and precise circuits to switch video signals into the mix at very high rates.

This entire grouping of modules is a *mix/effects bus*, a complete processing unit of switching components. There may be several buses cascaded together and used in a pyramid-like priority to create complex, multiple layers of effects.

The processing modules of the switcher are supported by additional modules dedicated to sync and video routing, distribution and tally indication (status display for operators). These various subsystems make up the basic architecture of the massive system we call the video switcher.

TIMING AND PHASING

All signals to be used by the switcher must arrive at its inputs simultaneously. Any delays between crosspoints will cause picture disturbances. To avoid this, some switchers will inhibit a switch to a source that is untimed.

To time signals to the switcher, H sync of switcher black is compared with each source. The source is then adjusted until coincidence is achieved. Failure to do this will produce horizontal shifting of the source. Likewise, V sync is timed to eliminate vertical shifting. Phasing of sources is the conforming of source subcarrier phase to that of switcher color black, sometimes called *black burst* or *crystal black*, to eliminate color shifts between sources.

The relative amplitudes of all signals are matched so that luminance and chrominance values are kept constant source-to-source. This is usually accomplished by comparison of all sources to a common reference, such as the *color bar test signal*.

SWITCHER TRANSITIONS

Cuts

The simplest and most familiar transition used in television is the *cut* or *take*. It is a switch transition in which crosspoints change during the vertical blanking interval. The switcher, like everything else in the facility, receives a feed from the sync generator via the pulse distribution amplifiers (DAs). The TAKE button on the switcher may be pushed at any time, but the switcher will not allow the actual crosspoint switch to take place until V sync arrives, a maximum of 1/60 of a second later. The "live splice" between the pulse trains of the two sources occurs in vertical blanking, where it is invisible. The active lines appearing on the screen after the switch are those from the second crosspoint, the B source.

Dissolves

The *dissolve* transition is not a switching action; it is a mix of the A source with the B source. Similar to the operation of an audio mixer, a dissolve consists of turning "up" the B source while turning "down" the A source. The result is a visual crossfade. A *fade* transistion is merely a dissolve to or from black.

Wipes

All other switcher effects are combinations or modifications of dissolves and cuts. A *wipe* transition, for example, is named for the appearance of the effect. In a wipe, the B source seems to "wipe" over the A source. A number of different wipe patterns are possible, from the simple to the complex. On the functional level, a wipe is rapid switching between the A and B sources outside of the blanking interval. Switching during active scanning is clearly visible. Unlike a cut, a wipe transits to the new frame at different places on different lines, and while cuts switch at the vertical field rate, wipes switch at rates higher than vertical.

Cuts always change the entire raster at once—wipes change the raster gradually, one line at a time. The different wipe patterns are produced by *function generators* —oscillators that can output a wide variety of differently shaped waveforms all phase-locked to sync. To produce patterns, these signals control the switching of the crosspoints at times corresponding to pixel positions along successive lines. The more complex the function waveform, the more complex the pattern of switches.

Keys

A family of switcher transitions called *keys* is similarly produced by active switching. (See Figure 4.2.) The *self key* is most familiar as a "super," in which text appears superimposed over the picture. The trigger or key signal, coming from a camera or an electronic character generator, is switched into the background picture at horizontal rate. The triggering of the switch is determined by the luminance level of the keying signal. When the key luminance crosses over a preset *clipping threshold*, the switcher cuts the signal on the line instead of the background. When key luminance falls below the clip, the background is switched back onto the line. This is why graphics on TV hot cards are printed white on a black background.

Figure 4.2: Keys

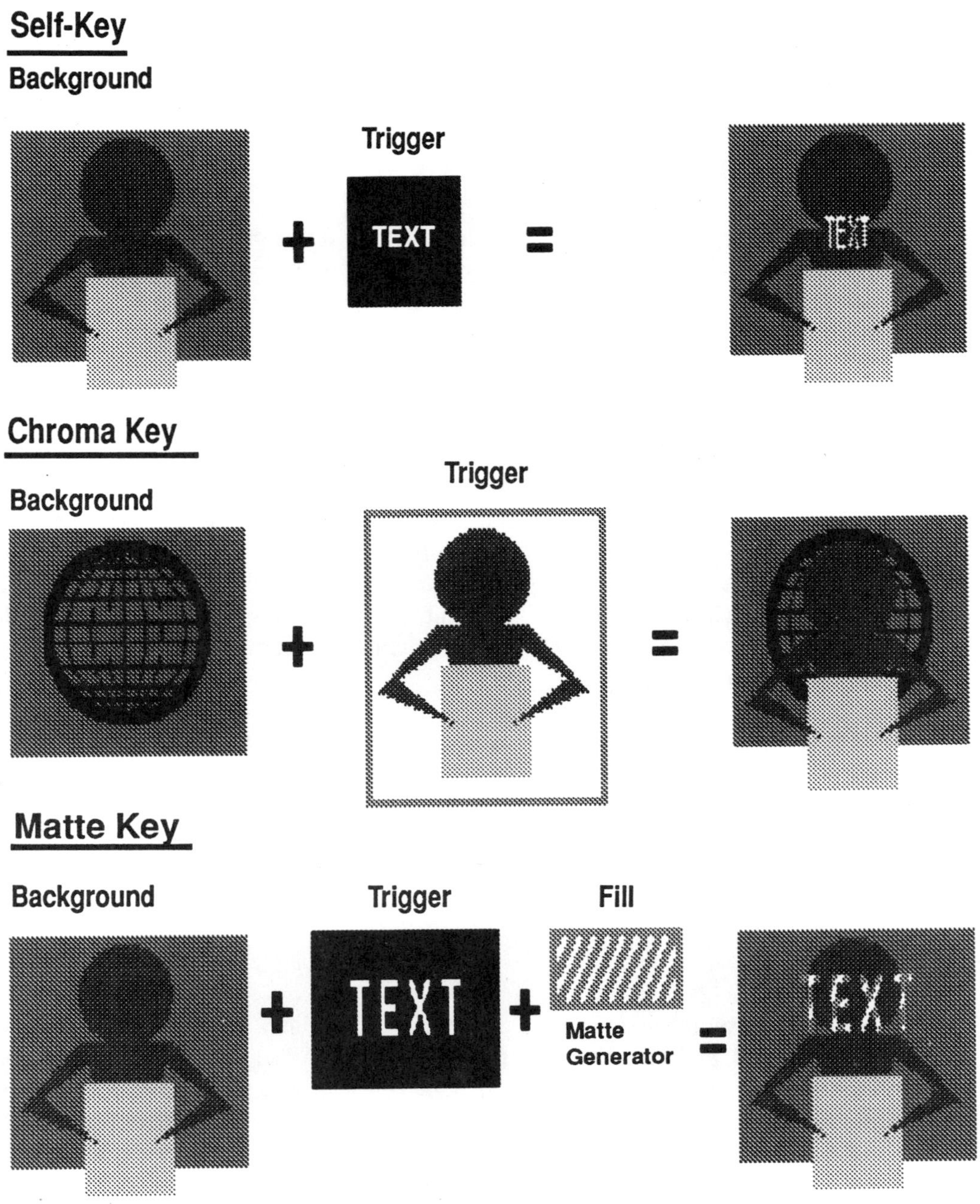

Chroma Keys

A variation on the self key is the *chroma key.* The same horizontal-rate switching occurs, but the trigger, instead of luminance, is the chrominance value of the keying signal. The chroma key is one of the external key family, in which the keying signal "cuts" a "hole" in the background picture. A third signal fills the hole. One of the most common uses for the chroma key is the satellite map apppearing behind the weatherperson on broadcast news shows.

What is actually behind the meteorologist is a plain blue flat that has been painted a specific hue known as "chroma key blue." Blue is usually chosen because skin has less blue in it than the other primary colors. This is also why weatherpeople are seldom seen on camera sporting blue clothing.

The maps and graphics in the background are usually generated by a graphics computer, or mounted on an easel and shot with another camera. In the control room, the technical director places the switcher in the chroma key mode, inserting the weatherperson as a key over the map/graphic shot. This key, however, is an inverted, or reverse, key which is active except where the blue flat shows up in the picture. The blue areas in the set act as a *hold-out matte*, allowing the graphic background to cut through, while permitting the meterologist to move "over" it freely. A hold-out matte is a signal which inhibits a key, preventing the effect from occurring in selected parts of the frame. Hold-outs are very useful when the "hole" moves through the frame.

Mattes

Another external key is the matte, in which text appearing as color-filled letters is keyed by a luminace signal, usually from a monochrome camera. The color fill is inserted from a matte generator, which is tunable to any hue, saturation and luminance combination. The matte generator can also be used to insert colored borders on wipes.

Using Two or More Keyers

When text is accompanied by drop shadows, borders or outlines, the effect is accomplished by two or more keyers working the same trigger signal; the shadow key is inserted after a slight delay.

Rather simple combinations of these basic functions can yield very elaborate-looking results. For example, a soft wipe is a switch from the A source to a very short dissolve (a fraction of a line) to the B source, followed by a period on the B source dictated by the function generator, followed by another microtimed dissolve back to the A source for the rest of the line.

THE SWITCHER CONSOLE

Earlier mention was made of the similarity between audio mixing consoles and the video switcher. In many respects, the theory of operation is identical. Video signals do require some different rules, however, which cannot be ignored. For instance, an audio mixer can fade a microphone out by simply turning the signal all the way down. A TV receiver requires a constant supply of sync signals to maintain a locked picture, so simply turning a camera signal down to zero will break up the display. A video mixer must fade to black by mixing timed and phased black signals into the fading camera signal. This is why there is always a crosspoint for black on the switcher, and why it makes good sense to time all signals to black as the reference—every show starts and ends there.

Switcher Design

The switcher is a good example of a machine that is designed to accomplish a set of tasks faster and more accurately than is possible for any human being. It is totally beyond the capacity of the human organism to observe, let alone supervise, the entire scope of a switcher's performance in real-time. A high level of automation is necessary for operation of such a device.

Nonetheless, the machine must be exceedingly "friendly" in its connection to the human operator in order to be useful at all. The objective is to maximize the information and control available to the human from the machine, while minimizing the confusion the person may experience because of the complexity of the system.

A technical director switching a live news program needs to know as much as possible about the status of the many options the show's director may call for at a moment's notice. This vital information must be provided to the technical director in a manner that is logical and easy to understand. It must be comprehensible at a quick glance and presented at a pace that allows for human intervention and correction. Above all, it must be supplied without undue distraction from the primary task—meaningful interaction with the director.

Switcher designers have always been caught in a delicate balancing act. As the range of possible effects has increased, the control interface for the operator has had to become simpler to deal with the expanded functions. The net result is always a more complicated control panel.

As complex as the control panel for a switcher appears to be, however, it represents the best efforts of those building it and using it to simplify the operation as much as possible.

THE TRIPLE-REENTRY DOWNSTREAM SWITCHER

Most switchers have control layouts that follow a basic pattern familiar to the technical director, although some controls will vary on different makes and models. Manufacturers may also vary on the terminology they use to label special functions.

The first level of organization is the *crosspoint,* the button used to select the input. The crosspoints are grouped into rows, the length of each depending on the number of inputs the facility plans to connect to the switcher. (See Figure 4.3.)

Two rows are grouped around the *fader* to allow selection of dissolve sources. The A row is the top row, the B row, the bottom. The fader is comprised of *potentiometers,* or gain controls. The fader is wired so that a single throw of the control turns one pot up while turning the other pot down the same amount. Sometimes it is desirable to independently control the fade of each side, so the fader may allow a "split," by which the A row may be mechanically decoupled from the B row.

The Mix/Effects Bus

The fader and its rows are a *mix/effects bus.* The bus also has a dedicated border/matte generator, a wipe generator, a keying amplifier and a joystick positioner, which is used to reposition wipe patterns. Different manufacturers may vary widely on the controls within the effects modules, since many options for combinational effects occur within these groups. For instance, soft wipes, bordered wipes and special patterns would appear as additional selection controls in the wipe generator group. The keying amplifier group would have controls enabling selection of mask-key and chroma key options.

During a broadcast a program director may call for a transition involving a scene that is itself an effect. An example is a dissolve from the weather set in chroma key to the news desk with a lower third graphic over the announcer. A bus is capable of performing only one effect at a time. A bus in chroma key cannot transit to any other scene until it exits chroma key mode.

To facilitate the transition of layers of effects, the switcher has multiple mix/effects (M/E) buses. This allows one bus to appear as a crosspoint in the next bus, thereby allowing a transition to a preset effect. The normal configuration is three M/E buses, labelled M/E 1, M/E 2 and M/E 3.

M/E 3 has additional crosspoints for the other two buses. M/E 2 has M/E 1 available to it. This is the triple reentry downstream configuration that allows any crosspoint to be layered three times in successive cascading through buses 1, 2 and 3. (See Figure 4.3.)

Note that a bus appears as a crosspoint only in a higher numbered bus. The technical director must work his/her way "downstream" during cascading, starting with bus 1.

The next layer downstream from M/E 3 is the program/preset bus, a bus without effects modules but with a fader and two rows. The program row is the "hot" row—the row that is on the air at all times. All crosspoints appear on these rows. It is here that upstream buses are selected or bypassed.

Figure 4.3: Triple Re-entry Switcher

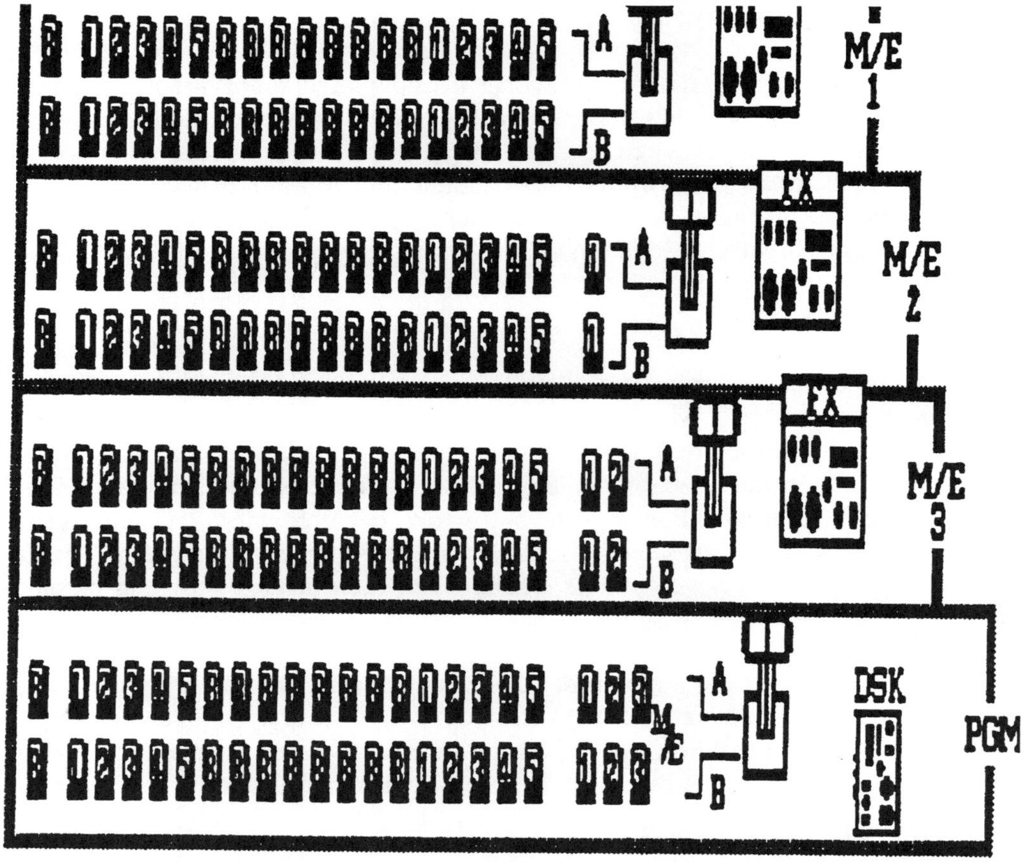

Downstream Grouping

Bus 1

Bus 1 + Bus 2

Bus 1 + Bus 2 + Bus 3

Bus 1 + Bus 2 + Bus 3 + Program

Downstream Fade/Key

Usually, there is a downstream keyer/fade-to-black module after the program/preset bus to allow a fade or super when the rest of the upstream buses are tied up.

SIGNAL PROCESSING

The switcher performs functions involving the mixing and switching of a wide variety of video signals. As previously mentioned, all input sources must be tightly timed and phased to a common switcher reference. Even so, some switcher actions, such as wipe and key, involve extremely rapid and complex processing of the inputs to produce the desired visual effect. If this processing is performed on *composite video* signals (those with sync pulses onboard), there is a danger that the pulses may be distorted, introducing time base errors into the output.

In the *non-composite* mode of processing, the switcher processes only the active video portion of the input signals. This is accomplished by passing the composite input through a clamping amplifier, which strips the sync from the signal and sends sync through a separate line to the switcher output (see Figure 4.4). This assures that the sync signals are unaffected by switcher actions.

At the switcher output, the sync signals are sent to a processing amplifier, which combines the newly processed active video signal with house or reference sync to maintain a properly timed sync pulse train.

Figure 4.4: Switcher Sync/Video Processing

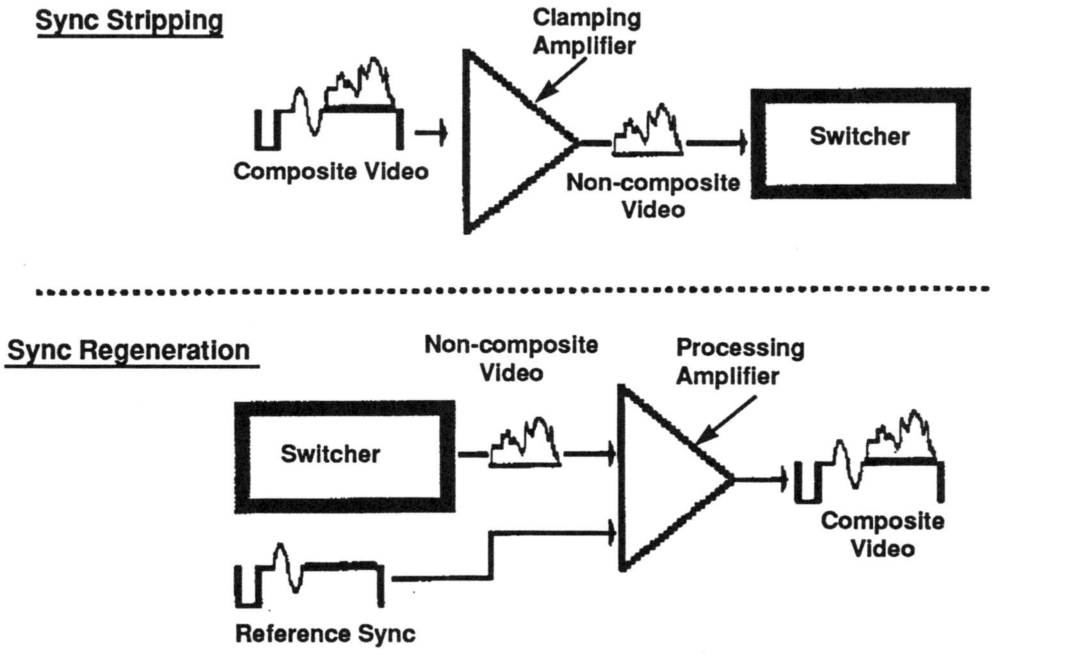

Because of the importance of maintaining the shape and timing of sync pulses, processing amplifiers (proc amps) are used liberally throughout television installations. Wherever a proc amp is used, fresh sync from the master generator is inserted in a video signal.

There are usually two distribution systems for routing signals through a facility: the video distribution wiring, which routes the entire composite video signal, and the pulse distribution wiring, which handles only sync. Of course, there are additional distribution systems for monitoring, audio and remote control.

GENLOCK AND THE FRAMESTORE

The sync generator is the timing reference for all signals processed through the switcher—ideally, all signals will be matched to it for seamless transitions. There are situations in which signals cannot be matched because they are far outside the range of local generator adjustment. Such a case arises, for instance, when a remote pickup (like a baseball game) is to be mixed with a local commercial or station identification. The cameras at the remote have their own sync generator, and the station's distance from the pickup constitutes an extremely long delay. Whether the signal arrives by land line (AT&T cable) or microwave (Studio-Transmitter-Link or STL), the signal travels miles further than that of the camera in the same building.

To reconcile this difference, the studio sync generator is slaved to the remote sync signals in the procedure called *genlock*. The local generator samples the incoming signals and retimes its output to match them, locking the local reference signals (in essence, the entire studio) to the remote generator. Although this achieves timing coincidence with the distant pickup, there is a big disadvantage. The master/slave relationship is backwards: the station is the more reliable reference, not the truck at the remote.

Any interruption of the remote transmission removes the reference signal at the studio, unlocking the local generator. If an STL suddenly goes down, the station goes with it. An errant technician at the telephone switching center could knock the entire station off the air by pulling one harmless-looking patchcord. Any fault coming from the remote truck spreads instantly all the way home.

The Digital Framestore

The arrival of the digital framestore corrected the master/slave priority. A framestore (also called a frameshaker) is a digital delay line coupled with a computer-controlled processing amplifier that has enough memory to hold an entire frame in storage. The computer is programmed to compare the incoming sync with station reference. The video frame is stored until the timing matches. It is then released to the proc amp. Output video is local sync with remote picture. The " frameshaker" made it easy to retime remote incoming signals to the local generator. First used during ABC network coverage of the 1976 Olympic Games, the framestore synchronized pictures arriving by satellite from the other side of the world, and has since been simplified and reduced in size and price.

THE MATRIX SWITCHER

It is hard to imagine "locking up" a triple reentry switcher until you are sitting in the "hot seat" as technical director or editor. "Lock up" is technical vernacular for "all buses are busy with current effects setups and the entire switcher is unable to make a smooth transit until a bus can be cleared." This condition of being painted into a corner with no exit is surprisingly easy to stumble into, even with three buses available.

The Grass Valley Group (GVG) and Central Dynamics (CD) have responded to this problem by introducing an expanded bus design for some of their upscale switchers, called the matrix bus. The design employs a digital assignment and control section for the fader and two additional rows for each bus (see Figure 4.5). A matrix bus also has two keying amplifiers.

The matrix bus is capable of transiting A to B backgrounds via the preset row, like normal two-row switchers, but it also can transit between either or both of two keys simultaneously. The digital control of the fader makes this possible. Before the fader handle is thrown, the technical director can select the row(s) to be affected by the transition. Any or all rows of the three possible selections, Key 1, Key 2 or background, will

Figure 4.5: Matrix M/E Bus

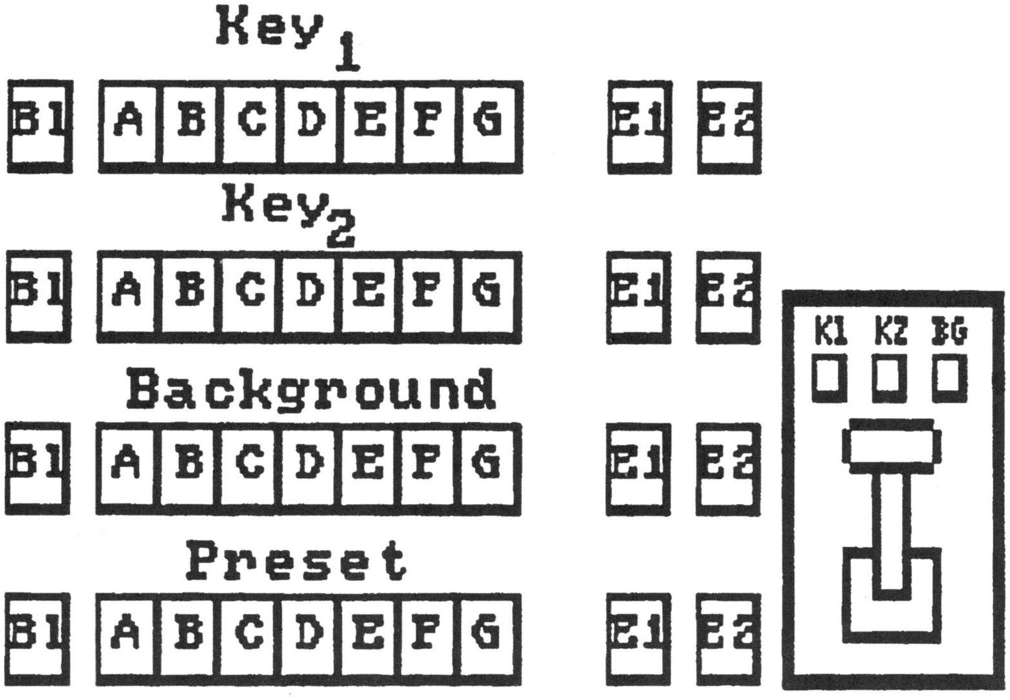

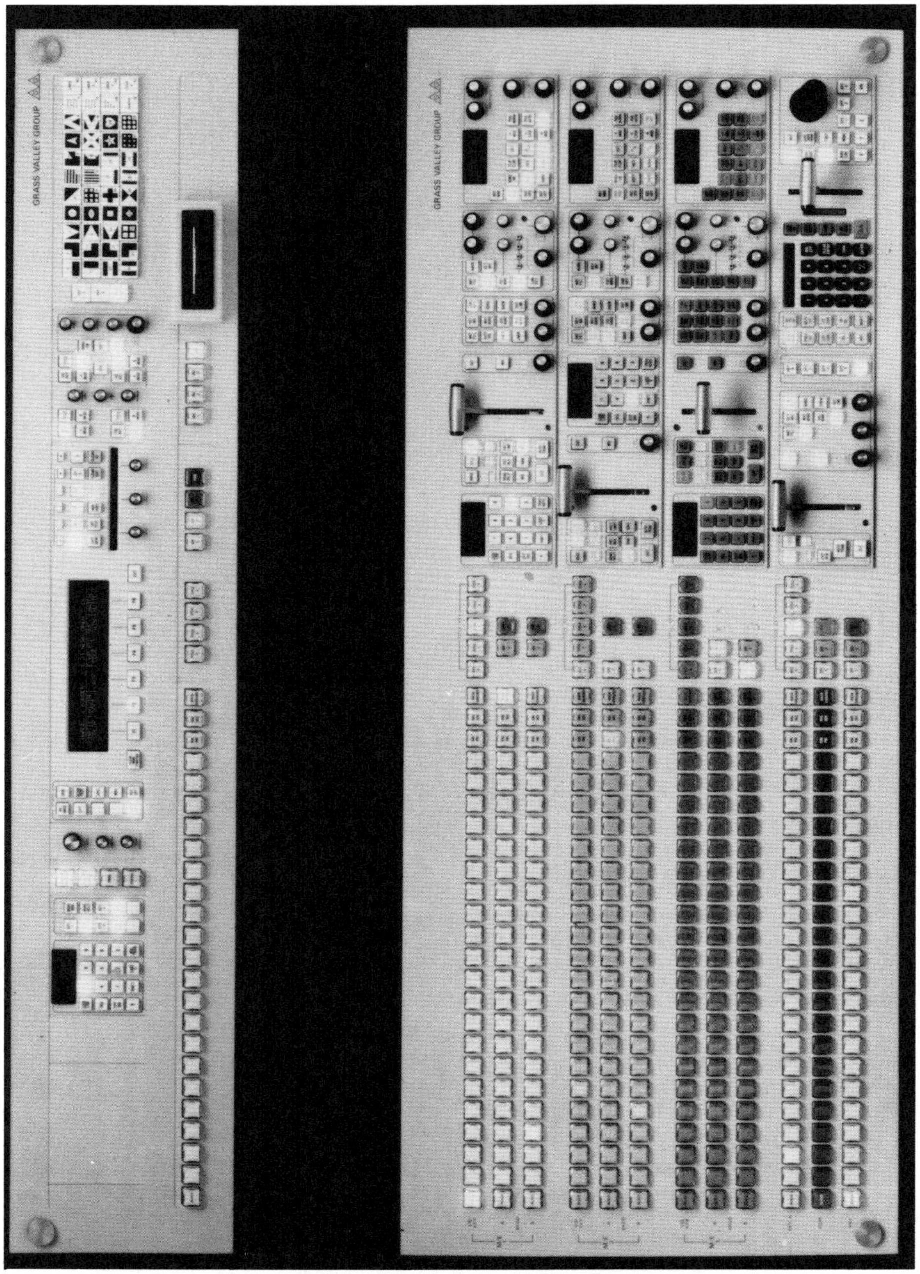

Figure 4.6: GVG Model 300-3A Video Switcher with Kaleidoscope™ Digital Effects System and Effects Memory System (EMEM®). Photo courtesy The Grass Valley Group, Inc.

transit if picked. If a row is not selected, it will be "held" in its original state through the transition. The added flexibility of assignable transitions with two keyers makes it easy for the technical director to set up layered effects and transitions on one bus where three buses were once needed. Central Dynamics uses this power to offer smaller switchers with a fewer number of mix/effects.

Grass Valley Group, on the other hand, takes a "bigger is better" approach. It offers matrix switchers of the 300 series in the multibus configuration, catering to customers who want ultimate switching power. (See Figure 4.6.) GVG adds another interesting twist to this luxury appeal: infinite reentry. There is no such thing as "downstream" in the 300 series: every M/E is a selectable crosspoint on all other M/Es, so the technical director can work "up" or "down" the tiers of M/E banks. This does *not* preclude the careful organization of switcher tasks to match the director's requirements—it is still all too possible to lock up even a three-bus 300.

If there is one rule regarding the role of the switcher in production it would seem to be—"there is never enough to satisfy the director."

INTEGRATED SWITCHING

Recently, switchers designed specifically for editing and automated production applications have appeared. These new designs depart dramatically from the traditional types. Sometimes, there isn't a fader or crosspoint to be seen; they are controlled directly by a host system's computer, allowing setups and transitional events to be loaded and executed through programmed automation. Such configurations are appearing in master control rooms and edit suites—the Grass Valley 1400 is an example.

As post-production technology continues to develop finer and tighter control, and as digital multitrack video recordings are used to advantage in editing, switchers will probably be scaled down in size and increased in control flexibility. Eventually, the switcher will probably be integrated completely into the next generations of production systems. The switcher will be controlled from the same terminal keyboard as the rest of the system, being nearly invisible itself.

These "virtual" switchers, which are more computer program than faders and wired switches, are here now, generating eye-popping effects. They are capable of storing setups and transition rates and executing them under fully automatic control with frame-accurate precision.

Nevertheless, the theory will remain constant: the timely switching and mixing of sources in matched coincidence make the magic of real-time effects and transitions happen. And the illusion will always have its roots in simple fact: the rabbit only looks like it comes from the hat.

5 Magnetic Recording

In Chapters 1 through 4, we considered NTSC television as a live broadcast medium. Although certain comparisons have been made to the mechanics of film along the way, television theory as a real-time transmission system has been our course of study. In this chapter, we will look at the fundamentals of magnetic recording and how they have been applied to the recording of video information.

RECORDING BEFORE TAPE

Television

During the first days of television, storing a program for later rebroadcast involved the process known as kinescope recording, which was nothing more than the filming of a monitor. The finished print could be televised by mounting it on a *film chain* or *telecine*—a projector optically coupled and electrically interlocked to a camera.

Radio

As World War II was drawing to a close, the primary method for recording radio broadcasts was electrical transcription (ET), a refinement of the Edison phonograph that would become the modern LP record. Although some tentative magnetic recording experiments using wire had been tried, the devices were primitive curiosities at best.

When the allies occupied Germany, they "liberated" a machine from the BASF factory; this magnetophone was a magnetic recorder, which used a ribbon of tape as the recording medium.

The U.S. Department of War turned the machine over to one of its prime contractors for study and further development. The contractor had provided the Army Air Force and Navy with sophisticated military instruments during the war and, though a small company, was recognized for its engineering expertise. The company's name was Ampex.

FUNDAMENTALS OF ELECTROMAGNETISM

Before we proceed with our discussion of the recording process, we should look at the fundamentals of electromagnetism.

When electrons move, two dynamic (changing) fields of energy are *always* observed accompanying them: an *electrical field*, the charge difference between the electrons and their surroundings, and a *magnetic field*, which acts like a magnet aligned at right angles to the electron movement. (See Figure 5.1.) As long as the electrons are moving, these fields are present. If the movement (current) ceases, both fields collapse. These fields are also directly proportional to the electrical current in both strength and polarity.

Electric and magnetic fields are interdependent: you can't have one without the other. A magnetic field arises when electricity flows, and a magnet in motion will produce a current. Electric motors and generators (dynamos) are examples of the latter.

The coupling of two different fields around a current can be used to measure and control that current.

Figure 5.1: Electromagnetism

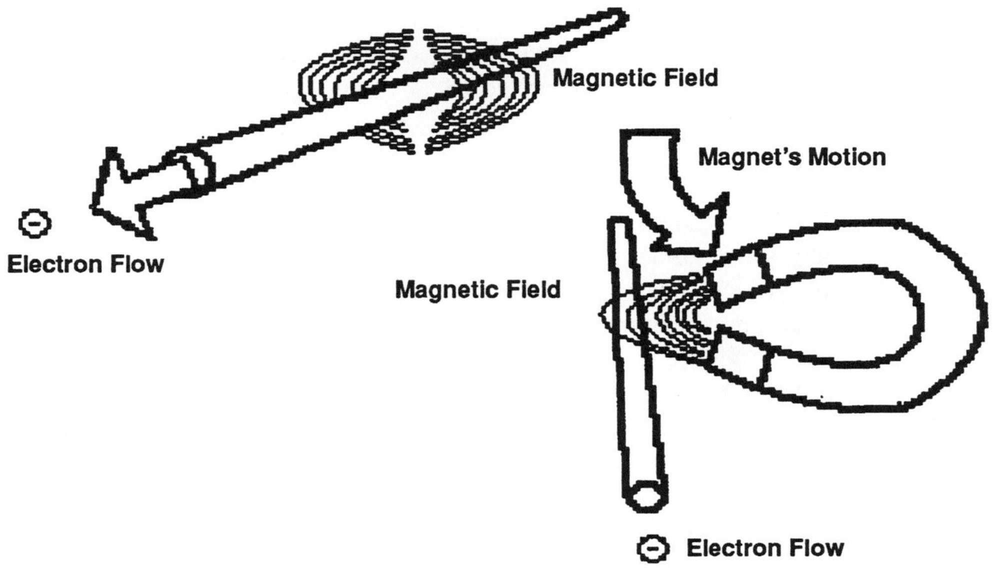

THE DIRECT RECORDING PROCESS

Exploiting electromagnetism for storage of signals involves the conversion of electrical signal wavelengths into lengths of magnetic tape. The first application of this technology was for audio signals. The audio frequency (AF) bandspread extends from 20 Hz to 20,000 Hz—the range of human hearing. A 20 kHz bandspread is 10 octaves. Each *octave* is eight full tones of music—*do re mi fa so la ti do*—a complete cycle on the musical scale. In engineering terms, an octave is the doubling of a frequency, so 40 Hz would represent the octave above 20 Hz.

The Microphone

The first task in recording is to convert the differential pressure waves (the acoustic energy) of a sound into electricity (an audio signal). The microphone is such an acoustic/electrical transducer; a transducer being a device which converts energy from one form into another. Modeled after the human ear, the microphone uses a pressure-sensitive diaphragm to sense the sound. To create an electrical signal that represents the sound, the diaphragm is attached to a magnet that is suspended in a coil of wire. As the diaphragm moves in response to varying pressure waves, the magnet moves across the turns of the coil and induces a current flow in the wire. This current (signal) is directly related to the amount and direction of movement of the magnet.

The Recording Head

To convert the electrical information into magnetic information, the signal is sent to another coil of wire, which generates a "phantom" electromagnet whose strength varies directly as the current varies. This electromagnet is the recording transducer, or head.

In the head, the coil is wrapped around a magnetically responsive iron core that is shaped like a horseshoe (see Figure 5.2). The core concentrates and focuses the magnetic field generated in the coil into a small but intense field at the gap—the open space in the horseshoe across which the tape is moved.

Magnetic Tape

To store this magnetic signal, a suitable medium is required. The medium must be magnetically "soft" enough to be "written" on, but magnetically "hard" enough to retain the magnetic record when the field is removed. It must also have a great enough number of *domains* (individual microscopic magnetic zones) to make a recording of acceptable detail.

The first magnetic tape was composed of a ribbon of acetate that acted as a base, or *substrate*. This substrate was coated with an iron oxide compound (rust) in a binder. Iron oxide was used because it responds to magnetic fields.

Figure 5.2: Magnetic Recording Head

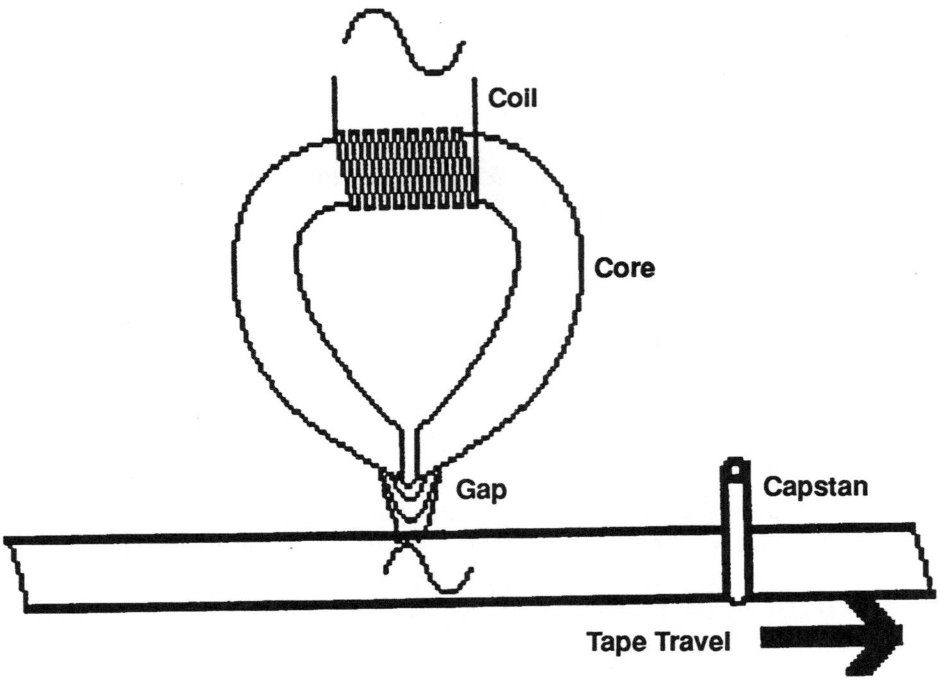

Since the tape holds the signal for storage, and the length of tape is the conversion factor for signal wavelength, the substrate must be stable while retaining enough flexibility to be manipulated by the recording machine. The tape should not stretch or shrink under dynamic tension, should resist breakage and be reasonably easy to repair if it does break. Acetate substrates performed quite well, but have been replaced with mylar base materials. These exhibit superior long-term storage properties because they are more resistant to heat and humidity damage. Mylar based tape is thinner, and longer loads can be packed on reels of standard diameters, thereby improving economy and storage space requirements. Unfortunately, mylar tends to stretch instead of breaking, which makes accidental damage to a master much more catastrophic than with the more brittle acetate bases. Modern tape recorders' motion-controlling servos minimize the likelihood of stretch damage, but accidents still happen, and many recording engineers long for the "good old days" when a clean break could be repaired with a piece of splicing tape. Nonetheless, acetate bases are unavailable now, and the only reliable insurance against stretch damage is the prompt protection of the master tape by an immediate back-up dub.

The Tape Recorder

The tape recorder itself is a machine with the electronics for electromagnetic conversion and a stable transport mechanism for moving the tape. The tape is mounted on a supply reel, moved past the recording head where the signal is written on it, and onto a takeup reel. The velocity of the tape as it passes the recording head determines the length of tape used to represent the electrical wavelength of the signal at the head. This velocity must be constant from head to tail of the reel, and must be repeated exactly when the tape is played back to prevent signal distortion. Simply using the takeup reel motor to pull the tape won't do, because the spooling of tape on its hub increases the effective diameter of the reel as the tape moves, and would cause the velocity of the tape to increase continuously.

To maintain constant tape velocity, a *capstan*, or precision-machined shaft, is used to drive the tape. The capstan pulls the tape from the supply reel past the heads. It is driven by a very stable motor at a precisely controlled, constant rate of speed. The takeup reel merely takes up the slack.

Playback

Once a recording is made, the tape may be passed by the same head again at the same velocity—*writing speed*—and the magnetic zones printed in the oxide will generate a fluctuating magnetic field at the head's gap. The coil will receive the field through core piece conduction, and electrons will move through the wire in response to the field strength. This is *playback*, the reading of the signals from the tape.

Remember that magnetic recording can only take place while fields are in motion. A static field cannot be recorded. The recorder reads the *changes* in fields as the information. If the tape stops, the signal stops. Similarly, if the tape's magnetic energy is uniform—unchanging—there is no signal, even while it moves. To erase a tape, merely remove the fluctuations.

Older recorder designs used a permanent magnet to erase the tape by imparting the same magnetic field strength to all zones. Newer designs employ a separate head fed by a high-frequency AC current. This produces cleaner erasures.

NON-LINEARITY IN MEDIA

The perfect magnetic medium would be linear in its transfer characteristics: for each unit of electrical energy input, it should ideally output a corresponding unit of magnetic energy. If graphed, this ideal behavior would plot as a straight line (see Figure 5.3). Unfortunately, magnetic materials do not behave so exactly—as a matter of fact, nothing does. A realistic plot of energy transfer characteristics shows marked curves at both ends of the range. This non-linear error is known as hysteresis. Lower electrical energies barely transfer at all, and higher energies tend to reverse the transfer characteristic. Between these inefficient extremes lies a short zone in which the desired linearity exists. It is here that the recorder operates best.

To place the activity of the machine within the linear part of this hysteresis curve, an AC bias current at supersonic frequency is mixed with the input signal going to the

Figure 5.3: Hysteresis

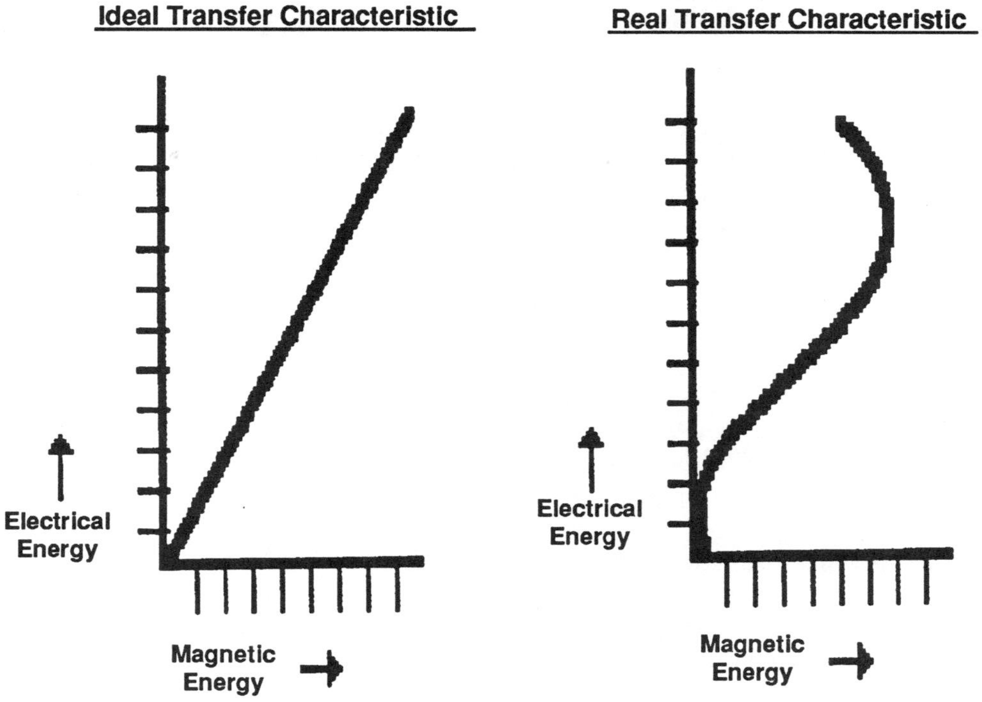

record head. This additional energy though transparent (not recordable), lifts the input signal above the low-end curve. The same principle is evident in the video black level of 7.5%. The phosphors on the CRT are not linear at 0%, so the elevated black acts as a bias for the picture tube. (Refer to Chapter 2.)

The upper curve is avoided by limiting the printing level on the tape—the voltage driving the head. Excursions of the signal above the *saturation point* (the beginning of the upper curve) overload the tape zones. Any additional signal after saturation has an erasing effect because of the overlapping of adjacent zone fields. Level meters on recorders are used to monitor signal voltages. The recording engineer calibrates the meters for accurate indications of recorder and tape performance. When they are accurately set up, the meters will display the signal levels and the operator can use them to reduce the amplitude of signals that might over-saturate the tape.

RECORD GAP LIMITATIONS

A magnetic recorder must be designed around the signal it is to record. Specifications for frequency response (the bandpass the machine records), dynamic range (the recordable voltages), signal-to-noise ratio (the difference between lowest signal voltage and the noise the system generates) and writing speed (tape-to-head velocity) are tailored specifically for the type of recording desired. As in all physical systems, no recording or recorder can be perfect.

One physical limitation is the inevitable imperfection in the machine's construction—for instance, of the head. The theoretically perfect head has a gap that measures exactly one half cycle of the input signal in width (see Figure 5.4). A gap with the perfect dimensions prints the maximum magnetic signal because the maximum amount of change is present in the gap at the instant of the printing.

Remember that the machine records signal changes, not voltages. If the gap is too wide, change is minimized due to cancellation of the different AC phases of the signal. This weakness is exploited in the bias and erase currents so the signals are not recorded.

If the gap is too narrow, the change state in the gap is less than is electrically present in the signal. This results in a printed signal with much lower amplitude.

Since it is possible to design a head with a perfect gap width for only a single frequency, most of the signals being printed will have errors introduced into them by the recording process itself. These distortions are amplitude errors as functions of frequencies outside the ideal gap width.

In audio recording, *most* of the signals representing audio tones are lower in gain on the tape than they were during recording.

Figure 5.4: Gap Width/Signal Strength

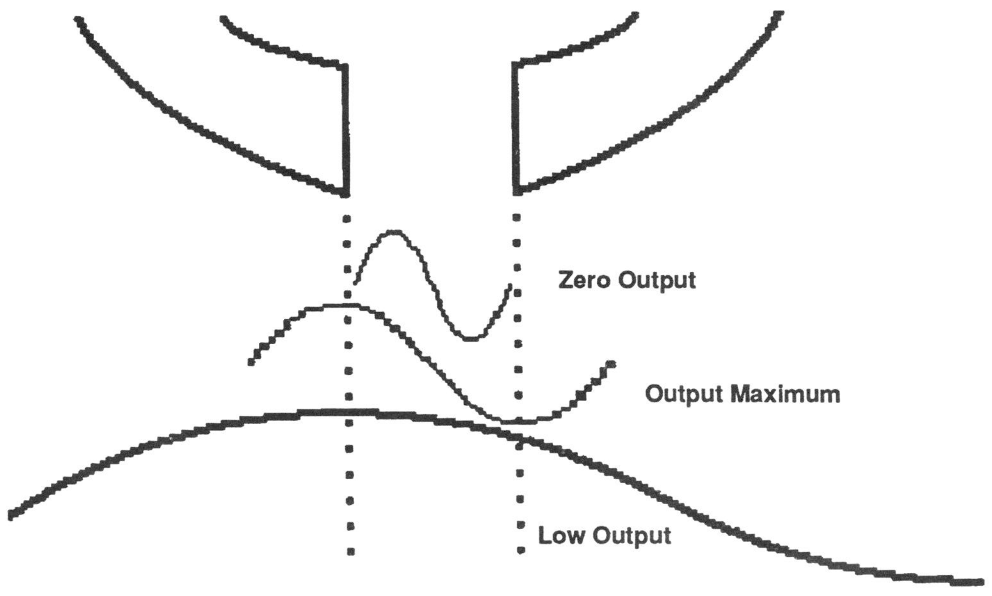

This condition is compensated for by *equalization*, in which input signals are passed through filters that boost amplitudes of the frequencies experiencing attenuation. Boost applied during the recording is called *pre-emphasis*. Most often, the non-linearity of the medium demands corrective filtering during playback as well, and this is *post-emphasis*. Because of the different behaviors of different oxide formulations, bias and equalization settings for a recorder may vary widely from load to load.

REQUIREMENTS FOR MAGNETIC VIDEO RECORDINGS

Many U.S. companies enjoyed the advantage afforded by the capture of the BASF machine. As a result, the manufacture of audio recorders improved and they were marketed worldwide. The d*irect longitudinal* recording format just described became standard for all applications in audio. This design features *stationary heads* that are precision-mounted on a rigid, machined chassis, which includes a capstan motor assembly for moving the tape perpendicularly across the heads. This arrangement produces longitudinal tracks of recorded information; the tracks are parallel to the long edges of the tape.

Tape widths and writing velocities were standardized to provide for machine-to-machine interchange of recorded programs.

Audio recording, with a 20 kHz bandspread, is done quite acceptably using this system. But the system is pushed to its limits at both ends of the audio spectrum and performs only with elaborate compensations to correct non-linearities.

The system's ability to make a recording is totally dependent upon the amount of information it can store on the tape. If a recording of a mid-band frequency audio signal, such as a 1 kHz tone at the standard velocity of 15 inches per second (ips) is microscopically examined, it would be seen that one cycle of the tone occupies 0.015 inch of tape length. In that length of tape, 600 particles, each representing a point on the wave, are available for imprinting. A good waveform recording is therefore feasible, even for a complex shape.

If a medium-range video signal is substituted, say 1 MHz, the wavelength of the signal is decreased 1000 times. The tape, moving at the same speed, now has 1/1000 of the length and particle density available for printing: 0.6 particle. No recording occurs because more tape is needed. If more tape space is moved during the cycle (higher writing speed), a recording might be made. The higher the frequency of the input, the higher the machine's writing speed must be. (See Figure 5.5.)

Another formidable obstacle to video recording is the greater bandspread of the signals. Audio is 10 octaves. Video starts at clamping, a DC value, and extends to 4.2 MHz; a range of 18 octaves. Throughout this range, the recording must have a degree of mechanical stability matching the electronic precision of the time-dependent video signals it will be expected to process.

Figure 5.5: Writing Speed

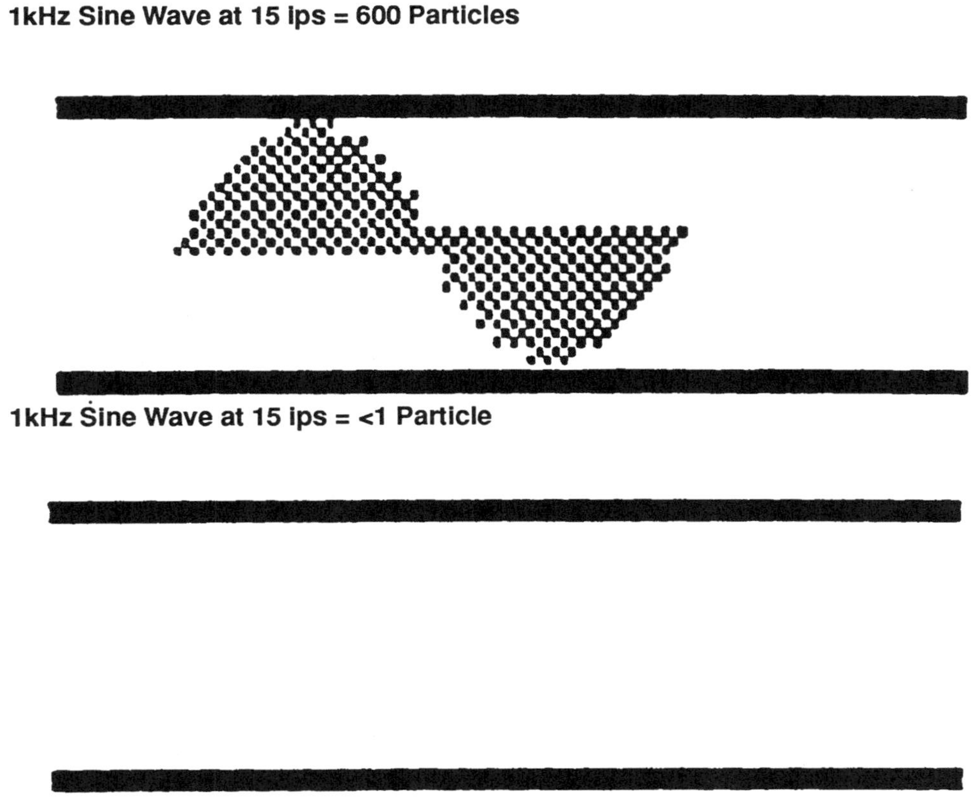

These conditions call for a degree of mechanical perfection throughout the manufacture and field operation of videotape recorders (VTRs) that was previously unimagined outside of research laboratories.

Development of Longitudinal VTRs

During the early 1950s, the Decca company, in England, and Bing Crosby Productions in the United States, had development teams working on prototype designs for *longitudinal* VTRs (LVTRs). Both companies demonstrated models that made monochrome recordings, but the performance of LVTRs was poor, and often, so beset with mechanical flaws, that the displayed picture could not maintain a lock for more than a few seconds. Because of the high velocity of the machine, non-standard tape loads were used. The capstan speed was 800 ips, making a five-minute reel 1/2 inch wide and 40 inches in diameter. Aside from the obvious mechanical difficulties in maintaining transport stability with such cumbersome loads, the cost of purchase and shipping of these awesome amounts of tape was a sobering prospect for producers.

THE QUADRUPLEX MACHINE

Meanwhile at Ampex, a three-man research team was taking a radically different look at the obstacles. These engineers realized right away that increasing velocity meant maximizing efficient tape use. Ideally, they wanted a machine that could use existing reels, since transport accuracy would be easier to guarantee if smaller loads were moving—tape economy being an important side benefit.

Including this goal of tape economy as part of the machine's intrinsic design was the key to Ampex's success. The *quadruplex* (Quad) machine became the first acknowledged standard for video recording worldwide and was the dominant format for more than 20 years. Although the quadruplex standard is considered obsolete today, it is more than a historical benchmark. Newer VTR designs still incorporate the same principles and theory that were developed for the Quad. As different as today's VTRs are, there is still much of the Quad beneath their controls. Understanding today's and tomorrow's VTRs is easiest when the original theory is clear. The ensuing study of Quad is useful because we are looking at more than an obsolete format—we are examining relevant engineering theory still very much in use today.

Transverse Scan

Having examined a number of different alternative track formats, the Ampex team concentrated its efforts on a transverse scan track for video, in which the information is imprinted at right angles to the direction of tape travel. Using tape two inches wide and 4500 feet in length (audio standard for 24-track), at a capstan velocity of 15 ips, a one-hour program would occupy a manageable reel of tape if enough tracks could be squeezed onto it.

To attain a high writing speed at a standard capstan speed, the video head would need to move across the tape, adding its velocity to the capstan's. Because the video tracks would be lateral (essentially vertical) instead of longitudinal (horizontal), the signal would be broken into segments during recording. These segments would be rejoined during playback to form a train of electrical pulses. The train had to appear unbroken, so the recorded picture would look as solid and continuous as a signal from a live camera.

The Quadruplex Head Assembly

The quadruplex head assembly was the design the Ampex team developed. In this design, four identical heads are mounted at 90° intervals on the periphery of a two-inch drum, or *headwheel*. The drum is rotated by a precision motor at a multiple of vertical rate: 60 x 4 = 240 revolutions per second, or 14,400 revolutions per minute (rpm). The track results for these parameters were: 3,452,400 individual two-inch tracks of a total length of 85 miles contained on a ribbon of tape 0.85 of a mile in length, with a net recording velocity of 1560 ips.

The small size of the rotating drum contributed to the stability of the machine, but the mechanical demands were still quite formidable.

Successful playback of the tape required precise centering of each video head on its assigned track, and the scanning of that track at exactly the same velocity that was used to record it. Each track would contain 18 lines of video, and the heads had to switch from track to track without any disturbing effect on the picture.

FM RECORDING

With track space at a minimum, and high head-to-tape velocity, there remained some complications. Direct transverse recording, even at these speeds, was still less than adequate. Aside from unacceptable frequency response, any irregularity in the tape path, such as a loss of optimum head-to-tape contact due to a dust particle, or wear on the head tips, would produce amplitude variations in signal output. These variations were displayed as brightness errors in the picture.

By using frequency modulation (FM), deviation of the carrier's frequency could represent luminance values over an 18 octave range. This technique is called frequency translation and is useful for extending the frequency response of recording systems. Desired bandspread is represented by "compressed" signal information, which is contained within the smallest actual bandpass that the machine is capable of recording. The system records an RF (radio frequency) envelope that is composed of the video signal and a 7.62-MHz carrier. Both of these signals are sent through a modulator during recording. Upon demodulation during playback, luminance values for the broader bandwidth are recovered and re-expanded uniformly as a function of the base carrier frequency. In other words, the *frequency* of the FM envelope represents the *amplitude* of the video signal—the picture's luminance. Since the recording's amplitude is less critical, the information is less prone to any voltage errors, and has a greater signal-to-noise ratio.

All video recordings are frequency modulated, regardless of head or tape format.

THE QUAD RECORDING CYCLE

In the Quad format, the video signal is presented to a modulator, which converts it into a narrow band RF envelope of the FM type. The envelope is amplified and sent through a commutator (rotary contact) to a video headwheel (drum) that rotates across a two-inch tape moving forward at 15 ips. The tape is shaped into a precisely curved "canoe" by a machined "shoe" with three grooves (vacuum guide). The shoe holds the tape in place against the rotating heads with a vacuum that is applied through two of the grooves to the back side of the tape. The third groove (relief groove) is a negative clearance provided for the heads at the arc over which they contact the tape; it penetrates slightly into the tape's oxide layer (see Figure 5.6).

The shoe forms the tape into a circular shape that conforms to the drum's outer rim

Figure 5.6: Quad Head and Threading Path

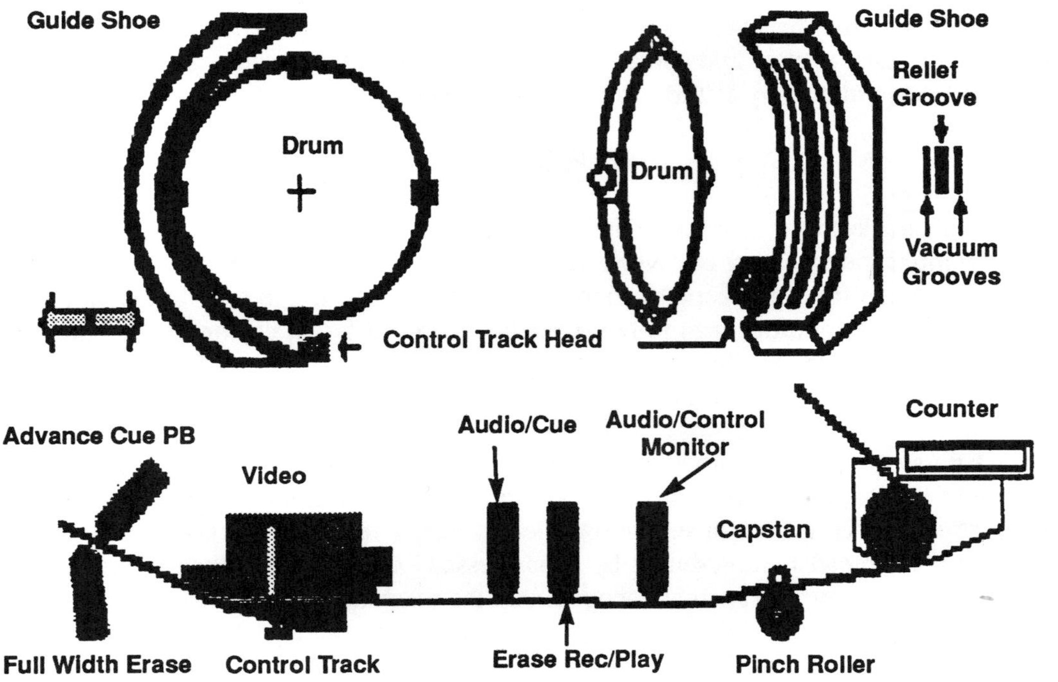

through 110° of arc. The four heads are mounted on the drum at 90° intervals. Each head's sweep overlaps those of the adjacent heads for 10° each, to produce tracks with redundant information at both ends. All four heads are fed the input envelope. The scanning action of the drum segments the tracks.

The Servo

The DC motor which powers drum rotation is fast-reacting and extremely stable. It is continuously monitored and corrected by a servo (error-detection) circuit (see Figure 5.7). The servo consists of a tachometer (tach) that is mounted on the motor shaft on which the drum turns. A magnetic or photodiode pickup senses the rate and phase of shaft rotation, and provides information on the velocity of rotation and tape-to-head contact at any given instant. The tach output goes to a comparator, where its frequency and phase are measured against the sync pulses on the video signal being recorded. If a difference is detected in the drum rotation as referenced to sync, the comparator generates an error signal. This signal offsets the motor drive amplifier current to bring the drum rotation into coincidence. This ingenius design effectively uses the timing pulses of the recorded signal to drive the scanning of the heads, assuring a time base for tape velocity that is, by nature, compatible with the input.

Figure 5.7: Quad Headwheel Servo

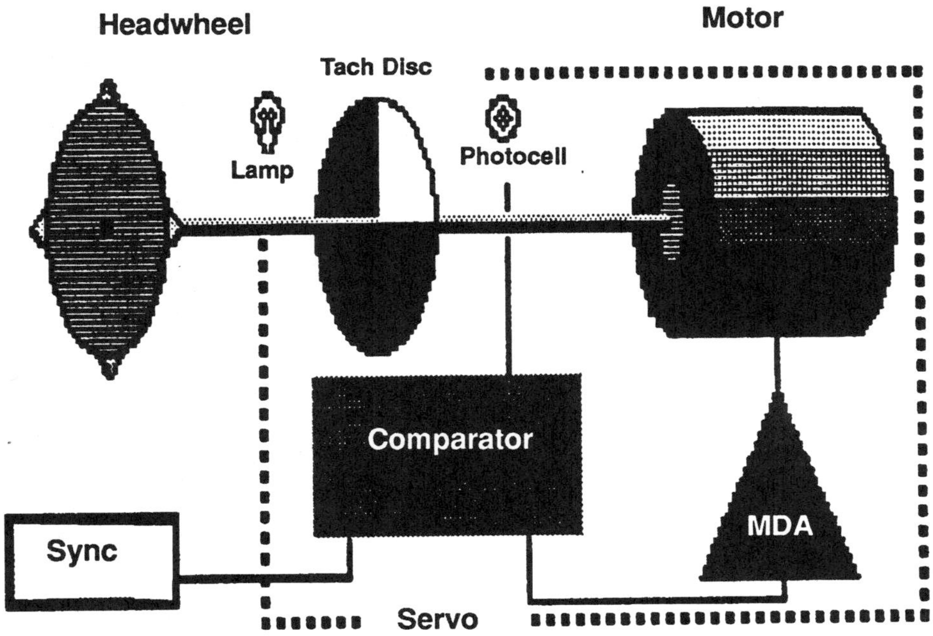

This type of servo—a velocity loop—provides fast and accurate feedback. Variations on this design are used throughout the different sub-assemblies of VTRs. Most of the automation that controls moving parts and electronic circuits uses some form of servo control.

As you might expect, the capstan velocity is also controlled with a servo motor loop. In this configuration, a summing amplifier produces a combined 240 Hz + 30 Hz control track composite waveform. (See Figure 5.8.)

Track Pattern and Time Base

The combined motions of the capstan and headwheel produce a track pattern of FM signals that are laid in very closely spaced, nearly vertical transverse scans across the center 1.82 inches of the tape. (See Figure 5.9.)

The FM envelope frequency varies during each scan representing luminance values along the 18.4 lines modulated for the 110° of rotation. The head velocity determines the time elapsed during the scan and, therefore, the number of lines included on the track—the horizontal time base of the display.

Figure 5.8: Quad Record

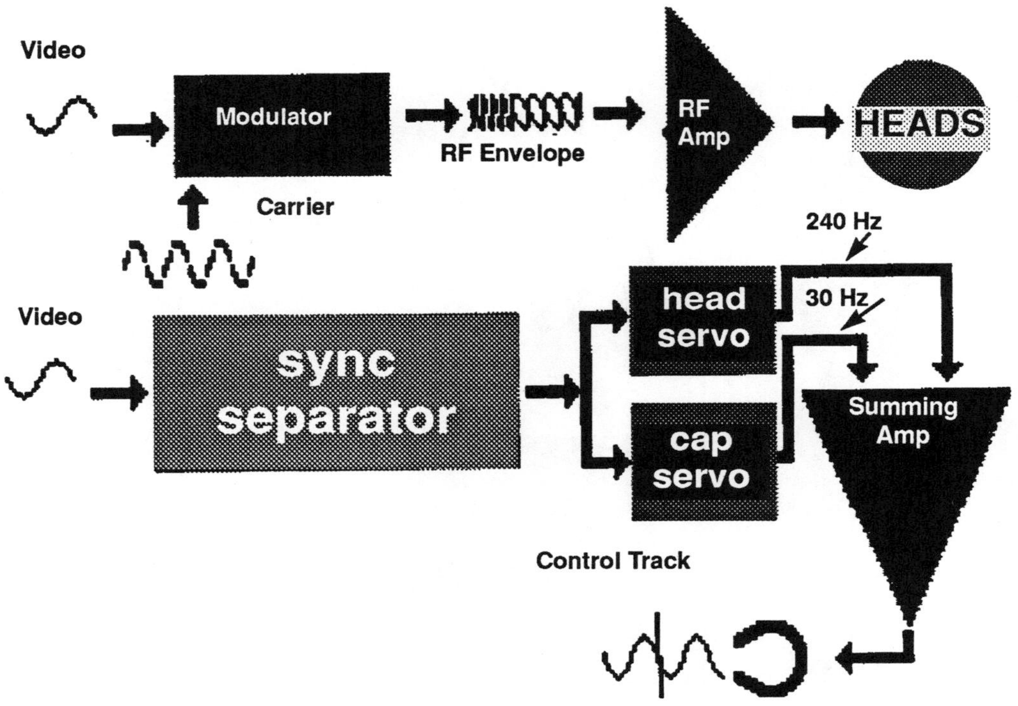

Figure 5.9: Quad Track Format

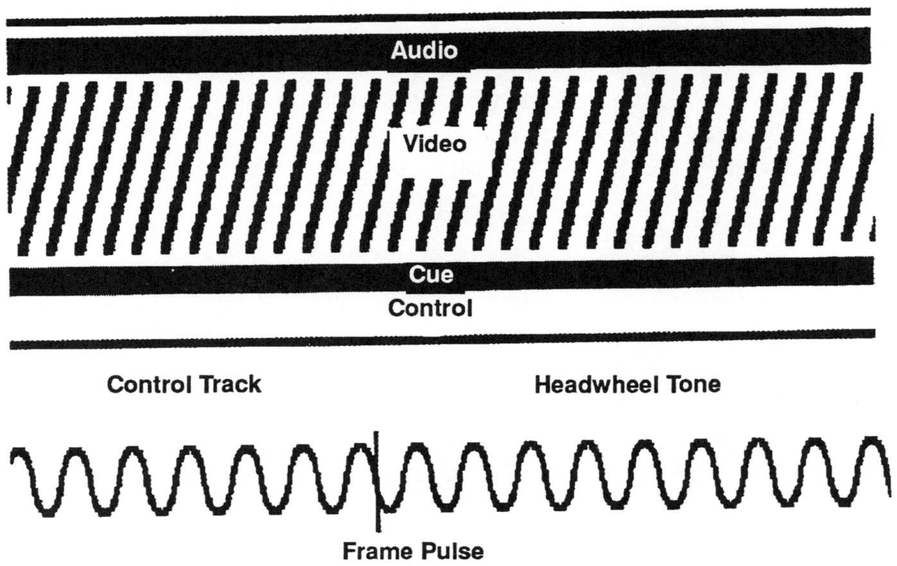

The spacing between tracks is determined by forward tape velocity and capstan rotation. This defines the stacking dimensions of each 18-line group—the veritcal time base of the display.

During one rotation, the four heads lay down 65.6 lines. In eight revolutions of the drum, 32 tracks are laid for one frame, occupying 1/2-inch of length. One line occupies 0.099 of one inch, scanned in 5.49° of physical rotation representing 63.5 millionths of a second of screen time. During that time the envelope can change 350 times to any one of 466 different values. A videotape recording is a monumental achievement in high-speed mechanics. Microsurgery is no more precise.

The other tracks on the periphery of the video scan contain AF band signals and are recorded in the conventional direct longitudinal method. In the Quad format, these are: program audio (top edge), cue audio (just below video) and control track (bottom, or reference edge). Each track is separated from the others by inclusion of a small empty space or *guard band.*

THE CONTROL TRACK

While the recording of so much tightly controlled information in so little space is certainly impressive, it succeeds *only* when it can be reproduced. Playback of such a complex format demands duplication of all the mechanical and electrical conditions that produced the recording. If a tape cannot be played back on a different machine of the same format, interchange compatibility is not realized. The range outside of theoretical perfection for system operating requirements (error tolerance) must allow a micro-accurate VTR to be treated like a lawnmower yet deliver broadcast-quality picture performance.

A Quad recording contains an astronomical number of segmented tracks, each one placed "on its end," and each corresponding to an exact number of lines on a display. During playback, the tracks must be unsegmented, that is, placed "end-to-end" again, in order and at the same rate, to simulate a live (not mechanically reproduced) TV signal at the output. The signal from the playing VTR should "look" to a switcher like the signal from a camera.

To do this, the VTR must be able to locate the center of the *home track,* a location on the tape where the *reference head* of the drum was positioned during recording. The Ampex standard defines head four as the reference head for the Quad format. When head four is at the center of each eighth pass across the tape, vertical sync pulses of the field division point (even-numbered lines) should be printing.

The position is specified by using the tachometer signals that were driving the motors during recording. The *headwheel tach* frequency (nominal 240-Hz sine wave at V rate) and phase are used to *position the wheel* with head four at 55° of arc. The *capstan tach* frequency (nominal 30 Hz at V rate) is used to *position the tape* to place the home track under the wheel as head four is crossing the 55° mark.

To make these tracking signals available for playback, they are combined and recorded as a composite *control track signal*, printed on the reference edge. The control track is a 240-Hz sine wave on which a 30-Hz pulse has been superimposed (refer to Figure 5.9). These are nominal values. If the tape has been stretched or the recording VTR's specifications are not completely up to format standards, the values will vary. But, they always vary in direct proportion to the track placement on the tape. The control track is a recording of the conditions under which the tracks were laid, as well as the current velocities required of the servos to recover the tracks, including variations in the tape itself after the recording was made.

THE PLAYBACK CYCLE

The VTR in playback mode first accelerates the headwheel and capstan to nominal velocities. When the PLAY button is pushed, a *pinch roller* is engaged from the back side of the tape, pressing it against the capstan. At this point, the control track is recovered and fed to the servos as a reference for the motor drives (see Figure 5.10). The servos lock up by finding the home track and phasing headwheel and capstan rotation to the control track signals. As this happens, the guide vacuum pushes the tape into intimate video head contact, and the tape RF signals are recovered. They are presented to the demodulator, where the carrier is subtracted, and a video signal remains.

Figure 5.10: Quad Playback Signal System and Servos

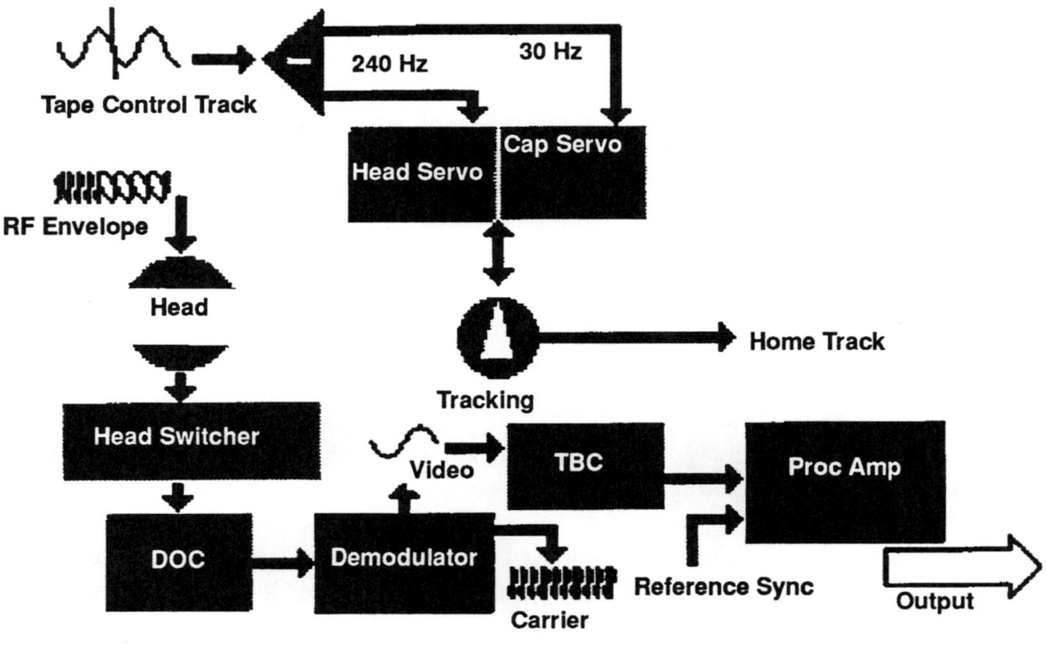

During recording, the video was presented to all four heads. During playback, only the head actually contacting the tape is turned on, thereby preventing noise present on unloaded heads from getting into the output. To unsegment the tracks, heads must be switched as they start or end a pass. Due to the 110° arc of tape used by heads every 90°, the beginning of the current track contains the same information as the end of the last track. This is a redundant switching period, and is approximately one line's length at each end of every track. Of the 18.4 lines recorded on each pass, only 16.4 are needed for playback.

A head switcher is coupled to the headwheel servo and switches heads during the redundant period to produce a continuous output. Switching action takes place during horizontal blanking to assure invisibility.

The video must pass through a proc amp, where reference sync is added in time coincidence with other sources at the video facility. This allows timing corrections for switching. The length of time required for a Quad VTR in good operating condition to track a tape and present a stable output is usually 5 to 10 seconds. Once a lock is achieved, the Quad demonstrates an awesome ability to maintain a solid picture.

When first demonstrated by Ampex at the 1956 National Association of Broadcasters (NAB) convention, in Chicago, the XVR 1000 MkI VTR, a bulky, expensive, noisy, hypersensitive, tube-laden monster with shrieking ball-bearing motor shafts and hissing high-pressure hydraulic lines, knocked the socks off the entire broadcast community and changed the way we would see the world forever.

TIME BASE STABILITY

Improvements to the Quad, quickly accepted as the new standard for TV recording, came fast and furious. The rickety ball-bearing supported headwheel assemblies were almost immediately replaced with frictionless air bearing heads, using pressurized air to support the motor shaft. Though introduced (like NTSC broadcasting) as monochrome, work started immediately on production of a color VTR.

Brilliant as it was, the original design left lots of room for improvement. Ampex came agonizingly close to theoretical perfection, but the stringent tolerances of television timing were well outside the physical capabilities of purely mechanical machines. Even air-bearing heads with solid-state (transistorized) servos were measurably short of the goal—VTR timing matching that of a camera.

Even with the tightest locks, the wheels, belts, pulleys and shafts betrayed their mechanical limitations when compared to scanning electron guns with no moving parts. For example, headwheels were routinely governed with enough accuracy to present a display to a monitor. But when a dissolve between a VTR and a camera was displayed, the VTR exhibited a noticeable horizontal jitter, the "chattering" of the bearings at the headwheel.

The VTR became a hybrid contraption. The velocity loop servos were capable of *time base stability* parameters adequate to lock the playback transparently within vertical scanning rates, but they could go no further. To tighten the timing accuracy to horizontal rates, a *time base corrector* (TBC) was added. The TBC was an electronic circuit consisting of an H sync comparator, an arithmetic averaging adder and a delay line of one-half H length. The jitter, mathematically merely a "blurring" of nominal line length, was calculated and averaged out by the delay line. The first TBC was called AMTEC, for Ampex Time Element Compensator.

Addition of the Color Subcarrier

By using this device, the stability of a system could be conformed to H rate and mixed like a camera—until the color subcarrier was added. At that point, color jitter became evident as pictures with hues wildly oscillating around their correct stable phase. Carrying the concept of time base correction further, Ampex added another option to the VTR, the COLORTEC (Color Time Element Compensator). The COLORTEC accessory narrowed timing correction to within 70 nanoseconds (ns) by extending circuit design of the AMTEC with shorter delays and faster comparator/averagers.

Ampex TBCs of this design worked remarkably well, so well in fact, that the servos were supplemented with a VELocity COMPensator accessory of similar all-electronic design.

The corrective ability of delay lines in stabilizing VTR output signals was established and extended. The DROPOUT COMPensator, a one-line delay inserted before demodulation of the RF envelope, was used to mask annoying flashes in pictures brought about by momentary head/tape contact loss.

The second-generation color Quads dominated the industry. Ampex and RCA shared the global broadcast market. But serious challenges from Japan's giants, notably Sony and JVC, were underway. These companies exploited the non-broadcast market, manufacturing smaller and less expensive VTRs for educational and institutional users. Sony's successful marketing and solid engineering was the start of what is now generally acknowledged as the complete takeover of the worldwide home electronics marketplace by Japanese interests. Today, they enjoy the lion's share of the professional audio and video markets as well.

THE HELICAL VTR

Sony pioneered the first non-Quad VTR format, the *helical scan* machine. The company developed the cassette VTR, its U-matic series, which was an overnight success. Soon, Sony found itself up against Ampex in the manufacture of a professional one-inch, reel-to-reel helical VTR—and Sony was winning. New designs have since proliferated that use different transports and tape formats, all based on the helical scan recording system.

The important distinction between Quad and helical formats is the video head. Helical video tracks are much longer than Quad tracks. A helical track contains as much as one entire field of video information. This is accomplished by enlarging the diameter of the drum and wrapping the tape around it during scanning. The tape enters the head assembly at the top of the horizontally mounted drum, and is guided around it in a corkscrew-shaped path—a helix. It exits the scanner at the bottom.

This method of scanning produces long diagonal tracks with more information and less segmentation than the Quad. The servo circuits are simplified, tape width reduced and cost (and so, price) of the machine is also reduced. The economy of this format makes it very attractive, though the signal quality is no better than Quad.

We will examine other aspects of the helical format in more detail in subsequent chapters. The physical arrangement of the tracks in longer scans is the big difference, which gave rise to a host of advantages over Quad, especially in cost-effectiveness and production applications.

Other aspects of the helical system are very similar to Quad: they both use FM modulation, servo-controlled motors, comparators and delay lines to restore signal time base, and processing amplifiers to restore sync pulses.

As the economics of the helical machines were becoming apparent to broadcasters, Quad VTRs were entering their third-generation design. RCA's AE-600 Quad sported an onboard computer for automatic editing and dubbing operations. The paragon of the Quad format was Ampex's AVR-1, which featured an all-digital signal buffer, a very long but finely tuned delay line, as a one-chassis all-purpose TBC. The AVR was built around a revolutionary transport design that used a vacuum capstan instead of a pinch roller. Its "intelligent" vacuum columns were equipped with photosensors and a microprocessor to guide the tape through the video head. These columns held three-foot loops of tape in reserve on either side of the head, while managing tension and contact during all tape movements. This isolation of the reels from the heads allowed the AVR to lock a tape into fully synchronous timing with live cameras in under 200 milliseconds, without a control track.

Though a marvel of engineering, the AVR was a marketing disaster. It was too expensive to make a profit. The Quad couldn't compete with the ruthless cost-effectiveness of helical VTRs. The third-generation VTRs were the last of the Quads. Ampex no longer manufactures even spare parts for them. The next machines, the Video Production Recorders (VPRs), ushered in the age of the *helical scan* or *slant-track* machines as primary recorders. Head design and tape format have undergone considerable change in the VPR, U-matic, VHS, Beta and 8mm evolutions of video recording. Yet the basic principles of electromagnetic transduction, FM compression and scanning transducers that convert time into tape length remain the same today. Perhaps the most remarkable aspect of today's videocassette recorders (VCRs) is how very much like lawnmowers they are treated, while performing as consistently as fine swiss watches.

6 Videotape Editing

The enormous success of videotape as a recording medium for television had an immediate impact on the film departments of networks and their affiliates. The traffic through telecine rooms decreased instantly. Because the videotape recorder (VTR) was intimately coupled to the television signal, it was nearly invisible in operation. From the very first Quad prototype, it was easy to mistake a recorded broadcast for a live one.

To use this to advantage in commercial production requires the ability to edit the footage. All editing, whether of film, text, sound or video, is the re-arrangement of the chronological order of events. Editing is nothing more (or less) than the movement of events in time, the shuffling of original actions or scenes into different relationships. The tradition of rivalry between editors of film and editors of videotape would seem to be totally groundless when viewed in the broad context of the definition: time manipulation. Editorial prowess is reflected in the ability to reorganize information; the medium in which an editor works is less important than his/her vision.

Before the invention of videotape in 1956, the only way to record a television broadcast was kinescope recording, in which a high-resolution TV screen was shot on film. The editorial method proceeded as a normal film process, completely within the familiar realm and control of the motion picture community.

While the image is *always* visible on film, it is only visible on videotape when that tape is being played back on a VTR. This would make a videotape editor of the early days more akin to a sound editor in the respect that they are both using opaque media. The difference between film editors and videotape editors has been the perceived difference of technicality required by the two media. The film editor has no less discipline than the tape editor. It is true, however, that the early videotape days were populated exclusively by engineers—the machines were incredibly complex to control by today's standards, teetering on the brink of breakdown every time a recording was made.

CUTTING VIDEOTAPE

Videotape has distinct divisions of each picture frame, although not as visibly as film. Like live television the videotape frame is divided at vertical blanking. Unlike film or live TV, videotape frames are "split" in editing at the halfway point. The odd/even division point—the *field division* corresponding to the junction of line 525 and line 2—is the "frame line" for conventional NTSC videotape. This is the traditional field two dominant editing standard as defined by Ampex and the Society of Motion Picture and Television Engineers (SMPTE).

The field two dominant position for framing is unique and recognizable as pulse train and tape tracks. It corresponds to the home track, the center of reference head rotation, which is marked by a 30-Hz frame pulse from the capstan servo during recording. If a recorded videotape is "developed," that is, coated with a suspension of fine iron particles, the magnetic tracks can be examined visually. *Edivue*, a solution of fine iron grounds in freon, was developed for this purpose.

In Quad, the video tracks are transverse, short vertical scans. In current helical (slant track) recording, the tracks are long and laid sideways. Quad's construction allowed razor-blade editing, while slant-track formats do not. Again, by following the evolution of the Quad, we can gain an understanding of the basic technology used by computer-assisted helicals today. The theory differs little among the different track formats.

Cutting and Splicing

Precision splicing blocks equipped with low-power microscopes allowed editors to position clips of developed videotape against optically calibrated scales and make razor-blade cuts referenced to visible frame pulses. These cuts bisected the home track at the third serration of vertical sync. When cut and joined (much like audiotape with an adhesive strip on the backing side), these clips became the edited master.

Editors wore absorbent cotton gloves while working to protect tape oxide surfaces from contamination with oils from their fingers. The cut point was located by rocking the tape by hand and listening for audio cues (a Quad recording displays a picture only when moving at locked velocities). After marking the back of the tape with a china marker and developing it, the editor would see a frame pulse—the cutting line. The clip was placed under the microscope and cut. The adjoining clip was positioned and trimmed to match, and a tab of splicing tape was positioned beneath the joint. After careful burnishing of the splice, the tape was lifted out of the microscope table and carefully washed to remove the Edivue. Only then could the edit be viewed. If the edit wasn't right, the whole process began again from step one, with one added hassle: the frames that had been cut out were lost. Cutting was grossly inaccurate in that the editor was virtually "flying blind" by guessing the edit without seeing it until it was done. Aside from the artistic frustration, there were physical liabilities (backaches) as well, because of the tedious nature of this method.

The Life Span of a Spliced Master

The edited master was a patchwork of splices. Each splice was a nearly vertical joint supported by a slice of thin-based adhesive tape. If the splice was good, only a small time base error occurred at the joint. This error was displayed as a "flag"—a slight horizontal displacement of the topmost few lines of the picture—and was usually not visible on the home TV set. (See Figure 6.1.)

Splices, when viewed as the physical result of cutting, were like scars or weak spots waiting to come apart again. All of these physical flaws had to pass through the Quad guide shoe and survive the murderous 85 mph head wheel, more than once. Each splice would go through the video head guide as a thicker, rougher section of tape, making the head "sing" as it scanned over the joint.

The dynamic loading of the tape during playback was also awesome: the reel motors were back-tensioned and pulled against each other to keep the tape tight and flat on either side of the vacuum guide. In this context, the video head was a circular saw "biting into" the topmost layer of oxide to recover the radio frequency (RF). Normal head wear and tape aging were exaggerated by the heavy contact and high velocities. Even under the best circumstances, each pass of the tape produced noise in the signal.

Dust in the tape path can physically scratch the tape. Splices were bigger surface features and almost always eventually destructive. The risk wasn't if a splice would fail, it was when. If a cut was good, it might survive five or six passes on a well-adjusted machine. On the other hand, a tiny error in thickness or symmetry of the joint could shred the master without warning.

Figure 6.1: Quad Splice

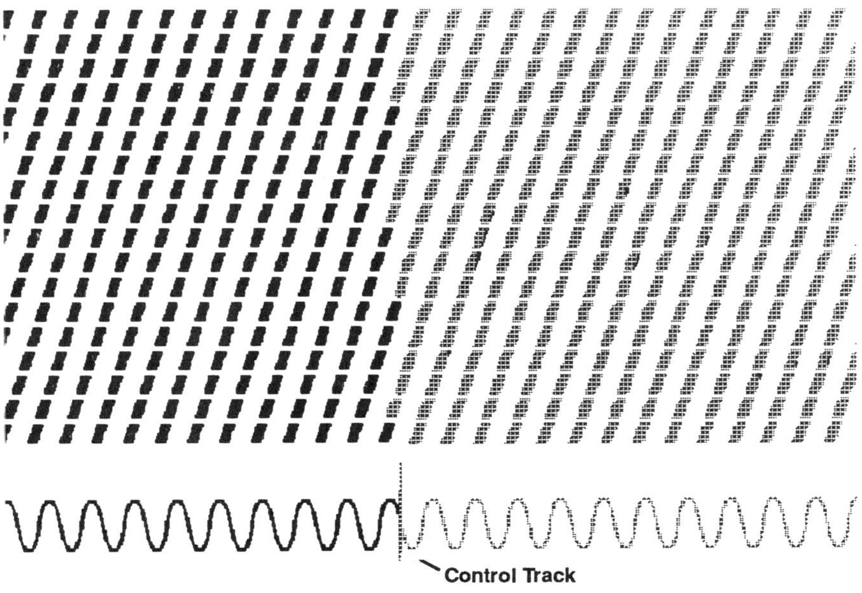

THE ELECTRONICALLY ASSEMBLED EDIT

Since the manually edited master could not be played many times after cutting, TV shows were *dubbed* (copied) for airing onto fresh uncut tape. The edited master was played only to make copies. Even then, the editor held his breath each time the master was played, awaiting the inevitable disintegration of the tape. Of course, this degradation of master quality happened gradually in smaller ways before the final catastrophe. A splice that was originally good would separate a bit each time it went through the head. Sometimes the joint would open enough to let the head scan the adhesive tape, thereby ruining a drum assembly. If the glue seeped through the underside of the cut, it would coat the tracks and unlock the servos, causing the picture to break up.

This crucial operation, which was routine in other media, had to be made more comfortable and accurate if videotape was to be accepted. The Mark I Electronic Editing accessory developed by Ampex was the starting point. The idea was to move the video tracks *electronically* instead of physically. While synchronizing the playback VTR to the recorder was relatively easy, erasing a single video track without affecting its neighbors was far more difficult.

To accomplish this task, a new composite stack of heads was mounted on the editing VTR. It held narrow-gap erase heads for the control track and video track. The capstan and head servos furnished all the necessary signals to trigger the single-track erase stack. The servos also synchronized the editing VTR to the playback VTR.

The Mark I was a simple electronic circuit that supervised the VTR's control logic, turning on heads at the proper times to simulate a physical splice. The editor would play the tape up to the cut point, and push the record button when he/she wanted to make a "splice." Using the control track as a reference, the Mark I would disable the video record mode until a frame pulse was aligned under the heads. When enabled, the record logic would be timed to energize the erase and recording heads in a sequence that spliced the signals being printed without cutting the tape. Not only was this "seamless" editing process easier on the editor, but the edits didn't flag, and the tape wasn't damaged.

In helical formats, a *flying erase head* is used to pre-clean the video tracks. The head is mounted on the video scanner ahead of the recording transducer and is energized exactly one track before the video record head.

Assemble Editing

Whenever a Quad VTR records on blank tape, a new control track recording accompanies the new video tracks. This is *assemble editing*, the simultaneous recording of new video and control track signals.

The leading erase head for the control track does not turn off as precisely as it turns on, however. The assemble edit begins quite accurately, but it ends rather sloppily. There are always some pulses missing (erased but not re-recorded) at the end of an assemble edit, and loss of the control track can break up the picture. Once a new control track is started, it must be undisturbed for the duration of the program.

To insert a new scene between two old scenes, physical splicing was the only way to guarantee that the *out* edit would be just as smooth as the *in* edit. Control track pulses had to be aligned and joined at both ends to form an uninterrupted signal.

THE ELECTRONIC INSERT EDIT

Of course, the best way to maintain a continuous control track is to avoid cutting it at all. If an edit involves *only* the video tracks, clean out-edits are possible as long as the new video tracks are correctly phased to the existing control track pulses.

The Mark II electronic editor was the next step. It was just like the Mark I in the *assemble edit* mode. In the *insert edit* mode, however, it "spliced" all tracks except the control track, using its phase to make a synchronous out-edit when the editor pushed the stop button. The VTR would continuously play the control track through the edit, using it to align drum and capstan rotation to simulate an "out" splice. The control track, being untouched, was likely to add even more stability to the EE (electronically edited) master's playback. (See Figure 6.2.)

With the development of the insert edit, editors could rehearse or preview a cut before committing to it. Enlarging the scope of delay line control logic, Ampex offered an accessory to the Mark II, the Editec (Edit Time Element Controller). Editec used an electronic cueing tone printed on the EE master's cue track to assure edit initiation and termination at electronically anchored points on the tape. To enhance the accuracy of this

Figure 6.2: Quad Electronic Insert Edit

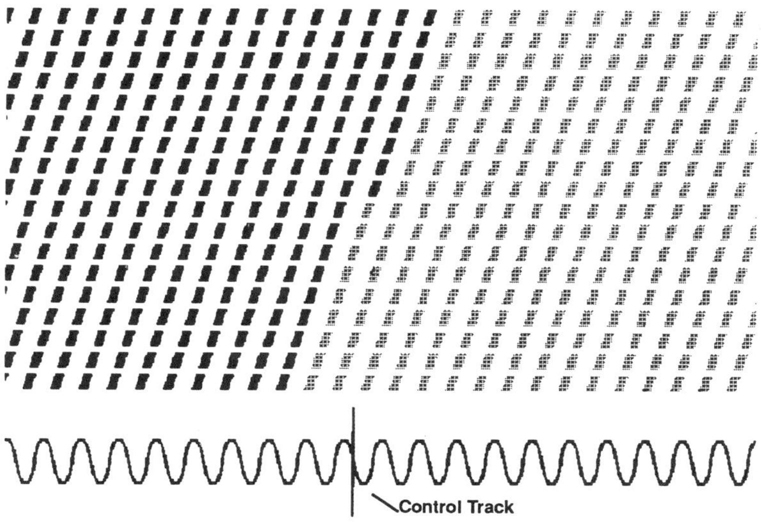

scheme, Ampex included a control for engaging delay line "taps" in single-frame increments. Once an edit cue was printed, the edit could be moved ahead of or behind the actual mark by engaging delays. To preview the edit, the editor played the edit-cued segment with Editec turned on. Editec switched from playback monitor to input and back again to simulate edits at the cue marks. If the editor wished to make the edit, he merely rolled the segment again, pushing the record button before the first Editec cue. This time, the VTR made edits based on the same cue marks.

The incredibly fine timing control afforded by Editec allowed videotape to be edited with all the accuracy of film cutting, in significantly more comfort, at greater speed and with none of the early drawbacks. In some cases, exotic editing (such as automated single-frame animation) was actually easier on videotape than in any other medium.

AUTOMATION IN EDITING

The development of the insert edit and Editec clearly put tape in the position of assuming the leading role in television production. All through this time, however, editors were still running from machine to machine, using china markers, cueing tapes by hand, gangrolling them and rewinding them separately. Video editing was definitely not a sedate, "sit down" kind of job. And film still had one major advantage that tape didn't—the edge number.

Each film foot had a unique label that made it possible to locate a frame in a reel as an offset from the nearest foot (edge or key) number. Negative cutters and optical laboratories routinely used this feature to locate and list key frames during film post-production. A certain degree of automation was afforded by having an "address" for frame location.

The first attempt to provide this information on tape was the "Lucy leader," an audio count recorded on the cue track. Desilu Studios, owned by Lucille Ball and Desi Arnaz, invented it. Using the Lucy leader, address information could be recovered manually and listed to an accuracy of about four frames.

The Electronic Editing Corporation (EECO), a "spinoff" group from Ampex, was interested in producing an address system for video by standardizing a machine-readable time and control code. EECO saw a future potential for the use of computers to automate VTR operations in editing and station programming.

EECO developed a computer, a program for editing and a code for addressing tape location. The EECO computer-assisted editing system and the Ampex RA-4000 computer-controller were released at about the same time (early 1960s). Both systems were primitive and unreliable by today's standards, but they shared one all-important central feature: a compatible machine-readable code that was recorded on the cue track and which gave each tape frame a unique address.

The Society of Motion Picture and Television Engineers (SMPTE) was invited to participate in the development of the code format to standardize it for all uses and users.

SMPTE TIME AND CONTROL CODE

The code that evolved is commonly called *time code* because its format is a 24-hour clock readout. It is produced as a series of rectangular pulses that are recorded on the cue track. There are 80 pulses per frame. The machine-readable code signal, a digital format recognized by computers, reads out as eight digits: hours, minutes, seconds and frames. Although the idea was to use the code both as an address for frames and as a program timing device, there was a complication.

Because of the addition of the color subcarrier, television framing was altered from the original NTSC specification of 60 Hz to 59.94 Hz (see Chapter 3: NTSC Color). The clock on the control room wall would run 0.06 of a second faster than the VTR frame code per field, adding up to 3.6 seconds per hour difference in elapsed time. The VTR framing could not be conformed simultaneously to the clock and the NTSC color standard.

SMPTE's solution to this dilemma was to create two different codes, one for time and the other for frames. The original code, now called *non-drop frame* or *full frame* code, retains all initial characteristics, including the time discrepancy. Non-drop code counts frames from zero to 29 for every second. The other code, called *drop-frame* or *skip-frame* code, eliminates the timing error by dropping two frame numbers each minute, except for every tenth minute. Drop-frame code counts frames from zero to 29 *only* on each tenth minute of show time. At all other times, it counts frames from 2 to 29, skipping frame numbers zero and one in the count. Though equipment may deviate from this rule (manufacturers take their own route in designs), the SMPTE standard is published as described above. Drop-frame is a time code; non-drop is a frame code. Drop frame code originally presented some problems to editors because of the missing numbers. But editing systems soon compensated for the eccentric numbering, ending delays and head-scratching caused by machines "looking" for numbers that weren't there. While there is still widespread use of both code formats, the drop-frame code seems to be the favorite of directors because of its timekeeping accuracy.

Time Code Phase

The code, regardless of format, must be aligned with the TV frame in the same way film edge numbers are. An address "straddling" frames serves no useful purpose; instead it generates an ambiguity. In order to assure accurate centering of each address on its frame, the code contains a *sync word*, a series of 12 non-address pulses that are phase-aligned to the videotape frame pulse and the signal's vertical sync at the frame line (see Figure 6.3). Code that is non-synchronous is not properly phased to indicate non-ambiguous addresses, and is generally unusable by automatic controllers. To phase-lock the code properly, the video signal's vertical sync is sent to the code generator while the recording is being made.

There is one other notable difference in code addresses as compared to film edge numbers. The edit point on videotape is at the half line division point in the frame—the middle, not the end. The code address does *not* increment at the match-frame edit point. A list of edit code addresses will show each "in" point at the same address as the last "out" point, not the next higher number, as in film. The cut occurs in the middle of the frame, not the end.

An *edit decision list* (EDL), in which each line represents one edit decision in machine-readable form, is illustrated in Figure 6.4. The listing is normal for computer-generated edit logs. So far there is no standard for such information. However, the different manufacturers of computer-assisted editing systems are aware that list interchangeability is a desirable attribute for users and are taking some steps in that direction. SMPTE is also pushing for a standard EDL format.

The first column of the EDL is the number of the edit line. Note that some edit lines (003 and 004 in Figure 6.4) are double lines, indicating that two functions are occurring as a single event. In this instance, match-frame cuts are made into the electronically edited master on the first edit line, and an automated transition is made on the second. The second column of the EDL indicates the source reel number. The third column indicates edit type: cut (audio/video), video-only or audio-only. And, the fourth column indicates the transition mode: cut, dissolve, wipe or key. Durations of transitions are shown in frames.

Figure 6.3: Time Code Phase

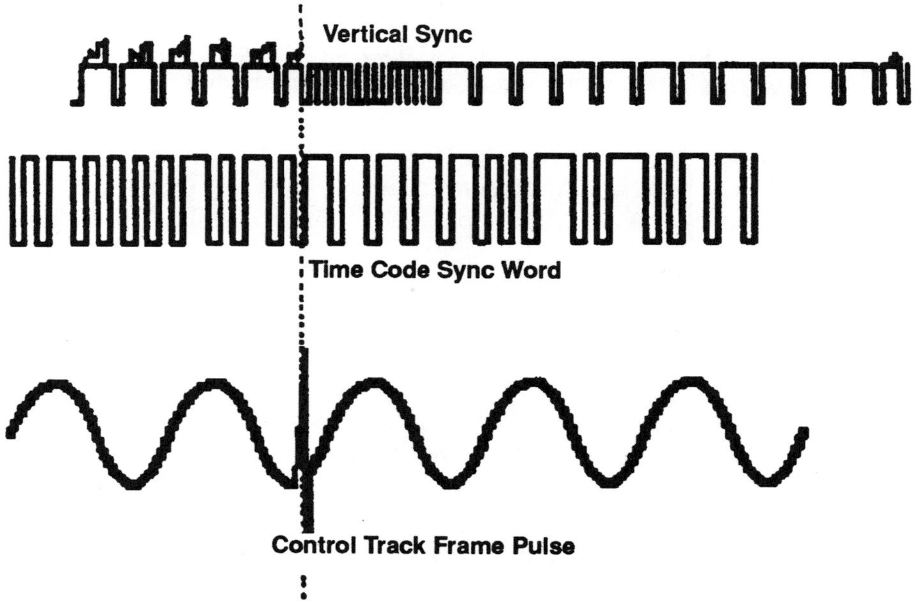

Figure 6.4: Edit Decision List

No.	REEL	TYPE	MODE	PLAYBACK IN	PLAYBACK OUT	RECORD IN	RECORD OUT
001	022	AV	C	01:00:01:22	01:00:02:00	01:00:00:00	01:00:00:08
002	011	A1	C	17:14:05:11	17:14:08:11	01:00:00:08	01:00:03:08
003	022	V	C	01:00:02:00	01:00:02:00	01:00:00:08	01:00:00:08
003	99B	V	D030	20:00:32:25	20:00:35:25	01:00:00:08	01:00:03:08
004	001	AV	C	09:22:02:10	09:23:02:10	01:00:03:08	01:01:03:08
004	BLK	AV	D015	00:00:00:00	00:00:00:15	01:01:03:08	01:01:03:23

The last four columns of the EDL define the edit placement as addresses on the playback (source) reel and the electronically edited master reel. In most systems, column four can be toggled to show either edit duration or edit out address.

Recording the Time Code

The code may be recorded in two ways:

1. as a longitudinal track containing the 80 bits as a broadband audio-frequency signal (longitudinal time code), or
2. as a digital video signal inserted in the vertical back porch of each field (vertical interval time code).

Longitudinal Time Code (LTC)

When stored as audio, SMPTE code appears on the Quad cue track, Track 2 of first-generation 3/4-inch U-matic VCRs, Track 3 of 1-inch SMPTE-C recorders or the "address track" of 800-format 3/4-inch recorders.

The advantages of LTC are its ability to de-couple the code and picture and record new code if desired. It also uses economical hardware to generate, detect, display and correct audio-frequency code signals.

Vertical Interval Time Code

Code in the video format, called Vertical Interval Time Code (VITC) is encoded as part of the picture. It is carried inside vertical blanking and is written and read by the video head assembly. The advantages of VITC are conservation of track space and the ability to read code reliably in still or slow-motion modes.

THE COMPUTER GOES NETWORK

After less than spectacular performance from the EECO and Ampex systems, CBS and Memorex saw the handwriting on the wall and formed a joint venture to develop and market a working computer editing system. The company formed from this alliance was CMX Systems. In 1970, CMX released its first prototype system, the CMX 600, to limited testing at CBS network flagship stations on both coasts of the United States.

In 1970, computers were very expensive to build and program—components and experienced personnel were rare. Though the CMX 600 tested well, the expense of Digital Equipment Corp.'s (DEC) PDP-11 minicomputers, and the cost of continuing software development and exotic peripherals like the light-pen terminal persuaded the partners that they could not afford to be in the computer systems business. They sold the CMX partnership, assets and patents to the Orrox Company, a silicon-valley based software distributor.

Bill Orr, president of Orrox, realized the media giants lacked some vital expertise. Though CBS knew broadcasting and Memorex produced recording tape, neither had experience with computers and software. Orr put a team of programmers to work revamping the 600 system from the ground up.

The Orrox team was aware that the future of computer editing hinged on economics and adaptability. The cost of the system would have to be attractive to a broad range of customers, and it would have to be flexible enough to operate in the different formats used by those customers—3/4-inch, Quad and 1-inch helical—in all possible combinations. The system would need to be just as easily (and quickly) updated as new developments in technology appeared. Doing all this *cheaply* was the trick.

Distributed Processing

The Orrox team isolated the main operating program and hardware (the "executive" or "big boss" of the system) from the "housekeeping" chores associated with controlling the different possible combinations of switchers, VTRs, audio mixing consoles and recorders.

Distributed processing was a breakthrough, not only in edit controller design, but in computer systems architecture (see Figure 6.5). The DEC PDP-11 minicomputer was used to run the main operating program. It acted as the "executive," managed the EDL, interacted with the editor, input, processed and output data, checked for errors and scheduled tasks. Instead of communicating with VTRs and other peripherals directly, the PDP-11 "talked" with an intelligent interface (I^2) that was connected to each machine.

The I^2 is a separate microcomputer programmed specifically for the type of device it controls. The I^2 "supervisors" acted on commands from the PDP-11 and were devoted entirely to controlling and reporting the status of their machines.

The CMX 300 system was the result of this modular system architecture. It relieved the "executive" from housekeeping chores, allowing it to concentrate on managing overall system performance. By reducing program size, it was possible to increase system speed. Conversely, the customer could change system configuration by updating an I^2, not the entire program, at great savings in time and cost.

Figure 6.5: Distributed Processing Architecture (developed by CMX)

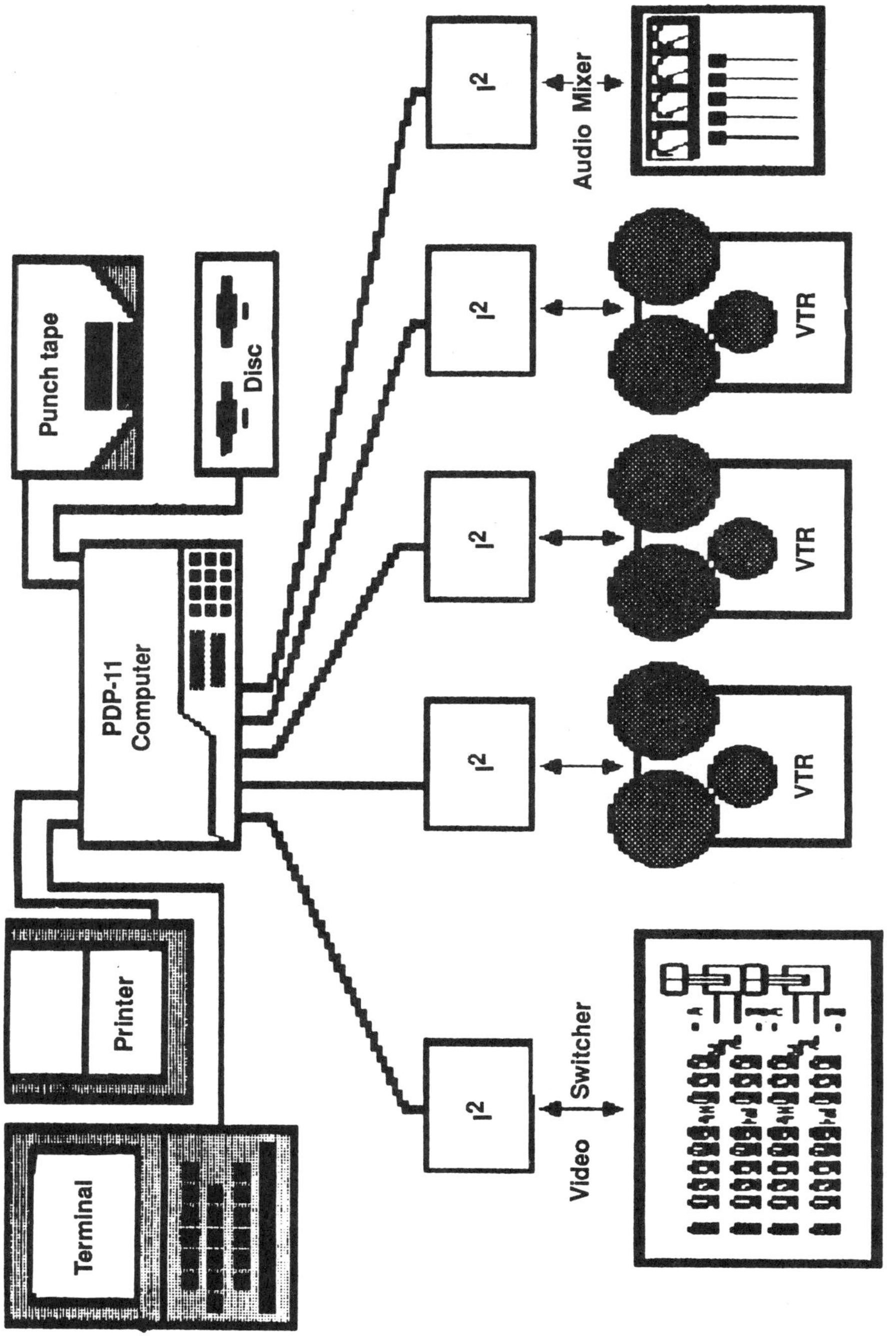

The CMX 300 system was designed to be economically feasible as well as technically sound. Because of its modular system design, the CMX has evolved continuously and economically through many updates: the CMX 50 3/4-inch system, the 340X all-mode system, the 3400, 3400+ and 3600 deluxe systems, the system 34 compact, the Edge economy-class system, the 6000 disc-based system and the CASS audio editing system. Other manufacturers that have produced editing automation similar in operation and design to the CMX models include GVG, Ampex and Sony.

OFFLINE EDITING

The cost of operating an editing facility, especially one equipped with automation, is extremely high. Even a small room with only three online VTRs, a switcher with limited automation, a "bare bones" computer and a small audio mixing console can cost over one million dollars to build. Many thousands of dollars in parts and manpower are necessary to operate and maintain such a facility each month.

This cost, when passed on to the user, can amount to several hundred dollars per hour in operating costs, which can cut deeply into any production budget. Automation can be used to offset and minimize this cost by maximizing efficient usage of a facility.

During the first stages of cutting, most of the time is spent reviewing alternative takes and deciding which works best against adjoining materials. It does not make good economic sense to use premium-cost equipment to do this when it can be avoided.

Fortunately, Ampex and a few other manufacturers—most notably Japanese giants such as Sony and Panasonic—had developed smaller-format VTRs for non-broadcast (industrial and medical) applications. These VTRs used tapes in widths ranging from 1/2-inch to 1-inch, and their servos were less critical than broadcast VTRs. This greatly reduced cost at the expense of picture quality, but it was a reasonable tradeoff for the non-broadcast market.

CMX and broadcasters availed themselves of these alternatives by interfacing the small-format VTRs with computers, and *offline editing* was born. In offline, edits can be previewed, recorded as rough cuts for director and producer approval and re-edited as many times as desired. This is all done on small formats for dramatically less cost. During the first offline cut, the computer is used to list each event, sort the events in chronological order and store the list. As editing progresses through successive cuts, the computer modifies the list to reflect changes in the show by inserting new events, deleting old events, re-sorting the list to accommodate the changes, recalculating the running times and storing the updated EDLs.

The Online Editing Bay

At the end of the offline edit, the computer dumps the list onto a punch tape, floppy diskette or other machine-readable medium. A hard-copy printout is also produced. The punch tape or diskette is then loaded into a "twin" computer in the online edit bay, where the master tape is assembled. The online system, which is equipped with the premium cost large-format VTRs, executes the edits that appear on the list. This process is called automatic assembly (auto-assembly). Though the term assembly is used, this editing is usually done in the insert mode on precoded, preblacked tape.

Assembly Modes

In auto-assembly, the EDL may be sorted by the computer to speed the assembly. In *A-Mode* assembly, the edits are executed in chronological order as they appear in the show. If there are many reel re-loads encountered, time may be saved by re-sorting the EDL for loading priority, called checkerboard or *B-Mode* assembly. In B-Mode, the reels are used throughout the show the first time they are mounted.

The computer may also re-sort the EDL by examining the playback source addresses and grouping the events that come on a reel together from the same or nearby addresses. This is C-Mode sorting, which can save a great deal of shuttling time in single-camera shoots.

The basic idea behind these different methods of sorting the EDL for assembly is to minimize the amount of time the system spends accessing the material. At $400 to $600 per hour, each second spent loading or shuttling a reel to bring up a shot is money down the drain. It has become embarrassing to load a reel more than once to get its shot into the show.

Computers are being used to great effect in increasing the efficiency of post-production.

THE SMPTE TYPE C 1-INCH FORMAT

The offline editing process fostered the growth of the 3/4-inch U-matic format of helical-scan or slant-track VCRs from Sony. The helical format, designed around a larger head assembly, called a *scanner*, crosses the tape at a much longer and more acute angle than Quad. The tape is wrapped around the scanner in a helical or corkscrew path that engages the scanning head at the top of the assembly and exits at the bottom. In helical recording, each video track is 263 lines long, as opposed to Quad length of 16 lines. (See Figure 6.6.)

The helical format is very popular for both offline (3/4-inch cassette) and online (1-inch reel) editing. One-inch helicals from Ampex (the VPR series) and Sony (the BVH series) have replaced Quad in most broadcast and production applications today.

Helical Scan VTRs

In the 1-inch helical format, the scanner supports three video heads: a flying erase

Figure 6.6: Helical Head Wraps

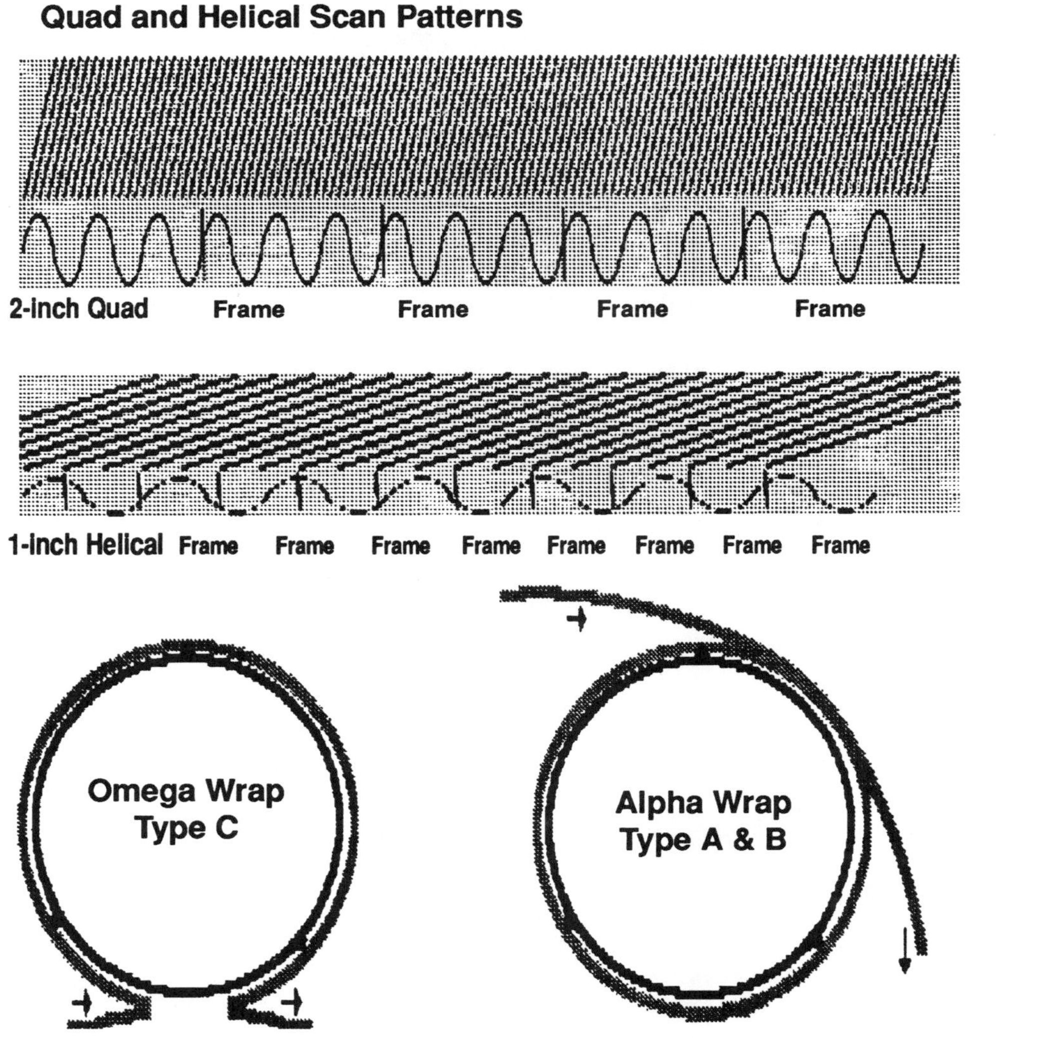

Figure 6.7: Helical Head Scanner/AST Detail

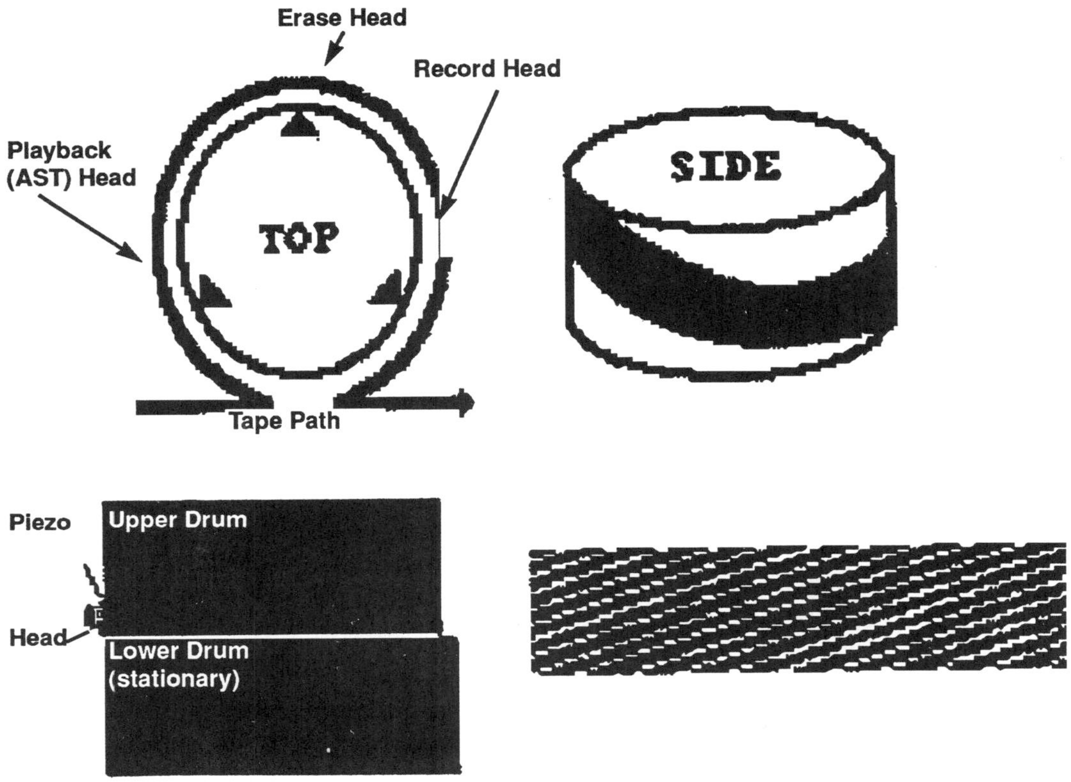

head, a record head and a playback head (see Figure 6.7). The heads are positioned 120° apart on the drum, which places them exactly one scan away from each other. The flying erase head can erase one video track at a time, leaving all other tracks untouched. Because of their positions immediately behind the erase head, the record and play heads scan the same track in order as the scanner rotates. The helical tracks are one field in length, eliminating the segmenting of active field signals into smaller sections. Thus, a SMPTE-C format helical VTR is capable of scanning a whole field in one rotation, yielding a "freeze" of the picture—a feature impossible on Quad. The separate play head allows the luxury of confidence playback, which is the ability to monitor the playback of the track while it is being recorded. The operator can see and measure the tape output during recording, confirming the quality of the recording without rewinding and spot-checking the tape.

Automatic Scan Tracking

Another feature unique to the helical 1-inch format is automatic scan tracking (AST) or dynamic tracking (DT). This feature increases the range of tracking during playback, allowing sub-standard or damaged tapes to be played. AST or DT involves a modification of the playback head assembly. The head is mounted on a specially built platform made of *piezoelectric crystal* (piezo-crystal), a material used in early "Marconi wireless" radio designs.

A piezo-crystal, when subjected to a current, will change its shape and dimensions in direct proportion to the amount and polarity of the current. This behavior is reversible: compressing a piezo-crystal will generate a current flow. Many "flintless" cigarette lighters use piezos in this way to generate a spark.

More flexible than the standard rigid playback head assembly, the AST piezo platform allows the head to reposition itself during a scan, changing its height to "hunt" for the center of the track. Using a servo loop (the AST servo) to sense the amplitude of recovered RF, any drop in signal level generates an error signal with amplitude and polarity representing a position off center. That error signal, when fed to the piezo-crystal platform, causes the crystal to change shape, bringing the scanning playback head to a new position closer to the center of the track. The fast action and accuracy of the AST head make possible the correct tracking of tape. This is possible not only during normal playback, but in variable speed play from still-frame all the way through fast-forward shuttle, and even in reverse. Pictures can be displayed *during* rewind and fast-forward, another impossibility on a Quad recorder.

The Benefits of Helical Design

The helical format also exhibited the benefits of greatly improved electronic circuit design and large-scale integration of micro-miniature components that reduced the size and bulk of broadcast VTRs. The helical could be set up on a desk. Compare this to the production models of the VR-1000 Quad machine, which occupied *three* racks each one measuring 19 inches wide by 2-1/2 feet deep by 6 feet tall.

The helical delivered astounding performance in this diminutive size, and it needed no control track to play a tape. Its signal output was so close to Quad in quality that measuring instruments were needed to detect the difference. Full editing capability was delivered as a standard feature, and it was simplified in operation. The multi-chassis TBC of the Quad generation was replaced with a single, small, digital TBC with a dramatically wider correction window that was virtually automatic in operation. The helical also featured three audio tracks for stereo program sound with time code separate as a standard feature.

THE OMEGA HEAD WRAP

The VPR and other SMPTE-C standard machines use a threading pattern for 1-inch videotape called the *omega wrap*. The path is shaped like the Greek letter omega (Ω), which causes the tape to contact the scanner for 346° of rotation. Three-quarter inch U-matic cassette VCRs and Bosch 1-inch VTRs use the *alpha wrap* (refer to Figure 6.6).

In the C-type omega wrap, the recording head is not in contact for a full revolution, so the entire video field is not recorded. The 14° gap at the flat end of the "horseshoe" is where the tape enters and exits the video head assembly. As the scanner moves the heads into that portion of the wrap, the information is scanned "into space" instead of onto tape because no tape is there to receive the signals. That arc consists of the last three serrations of vertical sync, the post equalization (eq) pulses and all but the last line of vertical back porch.

The AST and associated servos are instrumental in allowing the C-format VTR to virtually ignore this discontinuity in recovered signal. The TBC/proc amp section of the machine re-inserts new sync period and back porch signals from the house reference in proper timing relationship to assure a standard NTSC output from the VTR.

Ampex was justifiably proud in engineering a machine that could "break all the rules" and still produce an NTSC output. Nevertheless, some broadcasters were persistent in their demands for a machine, no matter how clever, that recorded a standard NTSC signal—ALL of it. Ampex was persuaded to include three more heads (sync heads) on the drum, offset 346° behind the video heads, to record and reproduce the missing part of the vertical blanking interval. Using these sync heads, the sync recording was switched into the playback during vertical blanking to recover a complete field.

When originally offered, the sync recording channel was an additional (and moderately expensive) option. Now, most SMPTE-C machines are sync-fitted as standard equipment.

EVOLUTION OF THE SMPTE-C FORMAT

The C-format has been steadily improved and updated through its many models. Notable among these is the Ampex battery-operated VPR-5 portable (see Figure 6.8), a machine remarkably similar in size and operation to the Nagra audio recorder, not surprising considering the cooperation of Stephan Kudelski (Nagra's inventor) in the design. The VPR-6 and the VPR 80 are full-featured machines with simplified modular construction and greatly reduced price tags. The VPR-3 is the top-flight version, featuring a microcomputer-supervised control logic function panel that automates all aspects of operation from recording setup to advanced multi-machine editing and animation.

Keeping pace with Ampex is Sony Corporation, which markets a full line of SMPTE-C VTRs with comparable features and quality, as well as a few that are unique in their capabilities. The Sony BVH-2500 Delta-T machine, with single-frame recording and playback capabilities, was designed with computer graphics animation in mind. The BVH-2800 Slo-Mo VTR, which can record at three times the normal NTSC rate for sophisticated variable motion effects, was designed specially for ABC network coverage of the 1984 Olympics. Also, Sony is the only manufacturer currently making a SMPTE-C VTR with stereo digital audio channels, the BVH-2000PCM.

Despite the undisputed supremacy of the C-format VTRs today, the future of the format has never been more uncertain. The same explosive technological development that fueled the rise of the helical VTR threatens its imminent overthrow. The newer *com-*

Figure 6.8: Ampex VPR-5 Portable Videotape Recorder

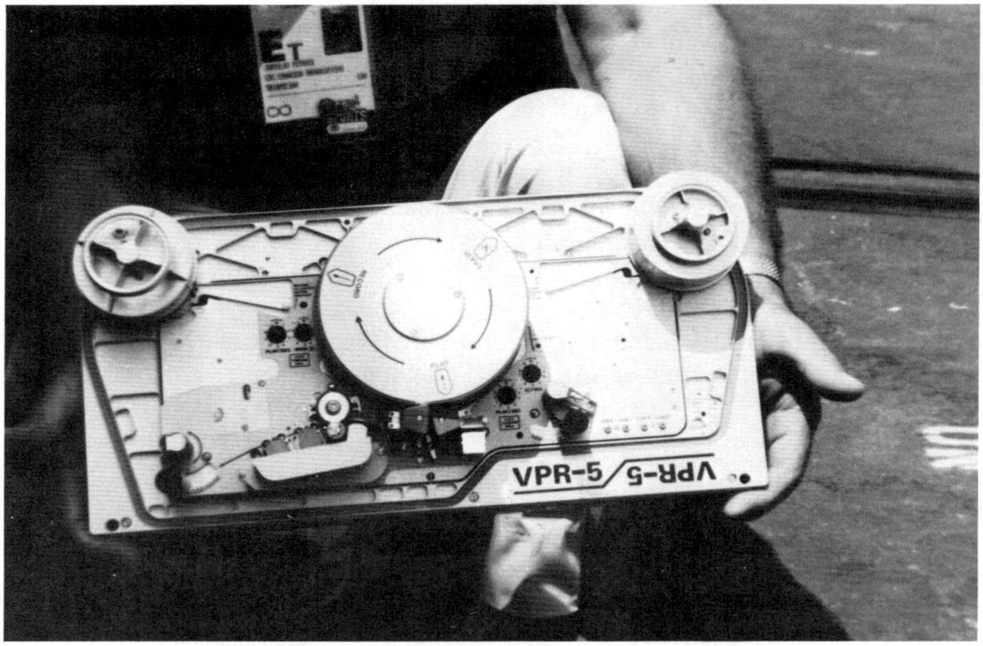

ponent designs (Betacam and MII Recam) yield superior quality on 1/2-inch tape, using cassette transports so reduced in size as to be incorporated into the camera itself. These designs record and process RGB signals using simpler and more transparent encoding schemes than NTSC.

As if this weren't enough of a threat, the biggest challenge to current tape designs, component VTRs included, is the undeniable ascendancy of an all-digital VTR, one that will make obsolete all current production and prototypical analogue models.

In 1988, Sony introduced a component (RGB) digital video recording format, standardized by SMPTE as the D1 VTR. It uses a small cassette transport and makes recordings that can be dubbed down hundreds of times without any detectable noise or signal loss. In fact, copies of D1 tapes are called "clones," rather than dubs. Clones, of course, are *exact* copies of originals. By recording and playing back the signal as RGB, the playback can be encoded into any international format. But, distributing RGB output throughout a facility calls for three times the wiring of an NTSC output, an expensive price for that quality and flexibility.

To counter such expense, Ampex followed up with a composite NTSC digital VTR of similar transport design, the SMPTE D2 format. Although not RGB, it shares the clonable attributes of digital recordings. And, of course, the wiring is cheaper. But the growing international distribution for video production raises a serious question as to the applicability of the D2 in the global marketplace.

As always, it seems the U.S. and Japanese manufacturers are pursuing their tradition of international competition.

But, that's not the end of the line. Another U.S. company with roots in the data-processing business, Abekas, is building recorders (the A-60 and the A-62) that don't use any tape at all! The Abekas recorders use Winchester disc packs (high-density computer storage devices) to record and play video signals in pure digital form. The Quantel Harry is another tapeless recorder which is used primarily for storing video processed through the Paintbox video effects system. (For further discussion of these systems, see Chapter 8: Digital Designs.)

NOVEL SYSTEM CONFIGURATIONS

Offline editing systems are evolving into many different configurations, "growing" in all directions outward from the hoped-for standard, all-purpose system. On the less costly side are the "micro" controllers from companies like Convergence Corp. These systems (the ECS-90, for example) lack the accuracy and consistency of the CMX and ISC controllers, but find a market by virtue of their greatly reduced cost. Unfortunately, the EDLs generated by these systems are sometimes difficult to translate into the data formats recognized by the larger systems, producing errors that often thwart the intended purpose—a smooth online auto-assembly.

On the more costly side are the "non-linear" offline systems such as the LucasFilm/Convergence EditDroid and the Montage. These systems use groups of microcomputers that run in tandem to keep track of a large number of playback machines, usually laserdisc players or Betamax VCRs.

Each of the 12 or so playback machines used by these upscale systems has a complete clip library of source material, and the EDL is really a play list of clips arranged in chronological order. The system takes advantage of the availability of 12 copies to "schedule" real-time chains of back-to-back previews of the clips listed. Thus, the search and cue of material on one source may take place while another source is in use, making changes in clip playback order virtually instantaneous, to execute and display.

The initial marketing appeal of this feature seemed irresistible. However, the cost of the necessary computing power, as well as the time and expense of making 12 transfers of all original material for each project, have given second thoughts to many of the original devotees.

An offline system that costs more to operate than the online system it "feeds" is economically questionable. Claims that these systems would be easier, faster and friendlier than more traditional computer-aided editing systems have faded. As of this writing, the Montage company has gone bankrupt (and resurfaced under new management with a redesigned product). LucasFilms has closed DroidWorks, its manufacturing subsidiary, and Convergence has merged with EECO and is redesigning its nonlinear systems.

Though some random-access systems still survive, notably the Spectra Image and EdiFlex systems, their applicability is considered limited, and their high cost remains a questionable investment.

Originally, non-linear editing systems were seen as a way to lure the film editing community into the computer age. Their proponents claimed that film people wouldn't need to learn about videotape to edit videotape. Editors lacking specific experience could lean on the computing power of the system, and the system would compensate. What appears to have happened is the opposite. Editors soon found themselves and their shows hopelessly mired in trouble when they hit online. The shows that are still edited on non-linear systems are cut by editors who know something about video, and who are somewhat acquainted with the systems they use.

The purpose of offline editing is the automation of the online assembly. The primary output from the offline process is an accurate list. The more accurate and complete the list, the more efficient the online session. Obviously, the best offline list comes from an editor who understands the online process.

CONCLUSION

Unfortunately, despite the continuing advance of TV technology, or perhaps *because* of it, there seems to be a reluctance on the part of many, particularly in the film community, to take maximum advantage of the evolutionary benefits. Although many traditionally based film editors have been compelled to "cut" a show on video for video, by and large they accomplish their tasks only with the aid of a battery of assistants who "mop up" after them. These assistants are eager and willing to help, but are being held back by the remnants of the hierarchy which still operates—a number of people who are unwilling to learn how to use the automation to its fullest potential.

The biggest mistake is to teach today's apprentices only yesterday's methods. It is unbelievable, but true to realize that most popular TV shows may be cut on a Montage or Ediflex, but involve just as much, or more, work than if they were cold-spliced on an upright Moviola. It is not unusual for a show's editor to have no clue about how an EDL relates to his/her show.

It's the same artificial boundary that many professionals declare when they ask if you are an "online" or "offline" editor. A good editor does it all. If you are editing video, you'd better know how video works.

7 Audio Sweetening

In this chapter, we will deal with the influence of television on sound. A comprehensive study of audio deserves much more space than we can afford it here—it is a book-length subject on its own. *The Electric Image* is concerned with the impact of the television environment on other technical arts and sciences. We therefore limit our examination of audio to the areas in which it connects with the video world.

DOUBLE-SYSTEM SYNC SOUND

Because videotape, like film, originally was physically spliced during picture editing, the tape sound track had to be handled in a fashion similar to film sound. In the case of film, the best way to register images is to support the raw film stock on both edges as it is exposed and moved through the camera's shutter gate. This is accomplished by the use of stock with double perforations. Double-perf stock has no room for a sound stripe, so a separate system is used to record sound. While the film is being cranked through the camera, an audiotape recorder captures the sound in a conventional longitudinal method. The audio recorder is fitted with an additional channel that simultaneously records a synchronizing track (Pilotone sync), which represents the rotational status of the camera's shutter drive motor. The Nagra from Kudelski is the most familiar example of such a high-performance battery-operated miniature audiotape recorder (ATR) capable of Pilotone sync operation.

Originally, this Pilotone track consisted of an AC voltage coming from a magnetic pickup on the camera motor shaft. The camera tachometer was connected by wire to the recorder and the tach signal was printed on the lower track of a two-track 1/4-inch tape. Variations on this basic method have appeared to enhance the process. These include center track FM, crystal sync, radio links, and now the introduction of SMPTE code on film. All of these, with the exception of code, recorded relatively long-duration motor tach pulse information. This information was useful only after an intermediate step called "syncing dailies," in which the editor positioned the tape and film at a common start mark on a gang-synchronizer. The common mark was made before each take by photographing and recording a "clapstick." Pilotone sync signals contain no address information and film edge numbers do not occur on each frame, so the sync marks were logged as foot/frame counts offset from the start of each reel (indicated by the "two-pop" cue on Academy Leader). The editor would manually sync and cut work prints of the camera negative for picture, and splice separate dubs of the recorded tapes for the sound, while carefully listening to each cut's location and duration.

In the case of videotape, the advantage of an onboard audio track was defeated by picture splicing. Two factors contributed to the waste: the large distance separating the synchronized recording position of audio and video tracks; and the transverse scan necessary for FM video tracks. Since this transverse scan was nearly vertical, the tape had to be spliced at right angles to the audio track, a bad angle for noiseless cutting. To cut pictures properly, the audio track had to be "trashed."

To restore the audio track, a separate recording and editing method was employed that was very similar to the film method. It used vertical sync instead of Pilotone as the reference. The process culminated in a synchronized *layback* of the cut, in which the mixed audio master was transferred to the edited video master.

AUDIO FOR VIDEO

The process described above, although extensively enhanced, improved and modified, is still employed today for productions requiring heavy augmentation or correction of the original audio.

With the advent of the Ampex Electronic Editor, physical cutting of the videotape became unnecessary. Editorial refinements realized by the Editec and then the minicomputer and microprocessor have liberated the editor and the tape from the domination of the razor blade.

Digital control systems and electronic editing allow a truly staggering degree of freedom in creativity, while preserving the physical integrity of the tape. Because of this flexibility, simple sound edits and mixes are routinely executed as part of the video editing process.

The Layover Process

More complicated audio processing is still generally reserved for the "sweetening" mixer. After the final video assembly is completed, the EE (electronically edited) master's audio and code is dubbed to a multi-track audiotape recorder during the *layover* phase. Usually a video dub with code "burned in" as video characters is made at the same time, so that a sync'd picture is available without subjecting the video master to wear and tear.

The Prelay

The next step is to assemble music, sound effects and additional dialogue tracks on the multi-track audio master, in sync with picture and production audio tracks. This is the *prelay* phase. The bigger multi-track recorders allow for many layers. Some productions are complex enough to warrant the ganging of two 24-track machines, involving 48 tracks in one mix.

The Mix

After all the tracks are in position, they are mixed down to a final, balanced set of composite audio tracks, usually on the same audio master. Sometimes the tracks are recorded on a separate four-track ATR, in which each of the four tracks represents a separate element of the final composite mix. These are called "three-stripe" masters, and are often used when a show is expected to be released in several different audio versions—two languages, for example. The tracks on a three-stripe are arranged as shown below, though the order of the elements may differ.

Track One	*Dialogue*
Track Two	*Effects*
Track Three	*Music*
Track Four	*Code* (sync information only)

The Layback

After the mixdown, the recording of the full mix back to the edited video master completes the audio post-production process.

AUDIO EQUIPMENT

Requirements for audio processing vary widely—no two shows are alike. Audio equipment has enjoyed the same beneficial influence from electronic innovation as video equipment and the impact has been truly astounding. In fact, in some respects, audio is more affected by recent developments than video, with the waves of change traveling faster and farther. This is due to the significantly lower cost (and lower risk) of implementing novel concepts in audio.

High-grade audio facilities capable of full-featured performance in a video environment are quite economically attractive. While this is also a trend in video, particularly in offline editing, the cost of achieving professional product specifications in audio is plummeting dramatically. While 24-track studios are the traditional norm for television sweetening, eight-track studios are quite common for audio, and are much less expensive to build and operate.

AUDIO LEADING THE WAY

The 20-kHz bandspread of the audio spectrum, when compared to the 4-MHz video spectrum, is immediately less demanding of untested designs, and affords a greater margin for experimental trial and error at the same cost. In most cases, and particularly in the instance of digital experimentation, what holds true for audio is true for video as well.

The audio industry enjoys a markedly wider application and user base than video: it is an indispensable element of television *and* film, as well as being the main attraction in the radio and record industries.

While it is apparent that the various media are crossing into each others' territory, it seems undeniable that audio has been a primary beneficiary of technological advancement, as well as the computer.

AN ORCHESTRA IN A BOX

Development of the Sound Synthesizer

In the early 1960s, a laboratory curiosity from the workbench of Robert Moog began steadily taking over the music industry. The Moog *sound synthesizer* was a monstrous collection of electrical circuitry that was able to mimic the sonic characteristics of traditional acoustic instruments, as well as create sounds never before heard.

The Moog synthesizer used a sophisticated system of voltage-controlled electronic modules to combine a number of simple signals into complex electrical waveforms.

The basic building block of the synthesizer was the *voltage controlled oscillator* (VCO). The frequency of the VCO's output was directly proportional to the amplitude of a voltage applied to its control input. The output signal was passed through a battery of voltage-controlled filters (VCFs) to modify the waveshape, and envelope generators and followers (voltage controlled amplifiers, or VCAs) to modify its dynamics. The signal was then mixed with other VCOs and noise generators to produce the final sound signal. A piano-type electrical keyboard was the controller for the many components generating and shaping signals. These artificially generated signals produced pleasing (for the time) sounds when they went through a loudspeaker—tones with musical attributes, such as *timbre*.

Moog's work paved the way for the establishment of the first standard for electronic music—one volt electrical/one octave acoustical change.

Over the last two decades, the synthesizer has been refined and improved and is a readily accepted, if not indispensable, part of our musical world.

Development of the Drum Machine

Back in 1967, the acceptance of electronic music was not so forthcoming. In England, the Roland Company, Britain's leading synthetic instrument manufacturer, produced one of the first *sequencers*. It was an analog memory (resistance/capacitance ladder) with a clock circuit which stored the keyboard voltages coming from the synthesizer while the musician played. When these pulses were later routed back to the synthesizer keyboard input, the sequencer "played" the tune exactly as the musician had. The sequencer was called the Roland Composer. Within a year, the Roland Company combined the Composer with a new product—another little black box equipped with digital memory. The memory contained recordings of single hits on various instruments in a complete drum kit. The analog sequencer circuits were replaced with higher capacity digital memory and clock modules, and the first drum machine was released to the public.

The reaction was overwhelming and shocking. Pop musicians and the experimental avant-garde loved the machine, grabbing up the inventory faster than Roland could replace it. The more established musical community, notably the musicians' unions, called for a ban on the device, stating that it would be used to replace live performances in concerts and recordings, thereby putting many musicians out of work. The controversy raged while Roland continued further research into employing the sequencer as a supplementary controller for all of its electronic instruments.

Public debate over musical automation was no more decisive than previous arguments had been during the industrial revolution. Finally, the rhetoric simply evaporated amidst the reality of the ever-increasing presence of fully supported electronic instrumentation. Though considerably muted, there are still disagreements expressed about musical automation. Some say that the "artificial" aid afforded by electronics degrades the quality of musicianship by letting anyone compose a tune without the benefit of study or musical discipline. Others say that the increased efficiency possible via electronic means enhances the quality of the final product by freeing the composer from mechanical limitations and supporting ultimate creativity. If one point in the debate is clearer today, it is that machines will never replace live performers, and a digital trumpet cannot ever sound quite the same as a brass one.

THE TAPELESS STUDIO

The drum machine inspired such hot debate because it was a totally new concept. It constituted nothing less than a solid-state tape recorder with no moving parts, which demonstrated a level of performance near to perfection.

Today, digital sampling combined with sequencing under the control of larger programs running on faster computers with more memory, enables musicians to electronically create much more than a song. Now, the instruments, the printed score—the entire studio environment—can reside on disc, be available at a keystroke, and be changed in microseconds. First seen in recording studios, these digital instruments are appearing in the sweetening rooms of video production facilities. They replace everything—the effects on cart, record players, tape machines, even the mixing console itself.

The E-Mu Systems Emulator, the Synclavier from New England Digital and the Fairlight CMI are digital sound workstations that are completely changing the way audio is handled in the record, film and television industries.

This powerful technology is even available in your own living room. Many personal computers (Macintosh, Atari, IBM and Amiga) support music and sound manipulation software, such as Mimetics' SoundScape programs, that come complete with multi-track recording and composing studios on floppy diskettes.

VIDEO FOR AUDIO

Audio has benefited greatly from the influence of video technology. The use of SMPTE time code in sound recording and mixing is a direct legacy of video. Computer-assisted automatic mixing consoles, such as the Solid State Logic 6000 series from SSL, make extensive use of video display terminals to enhance operation, and are designed expressly for use in the video environment.

Pulse Code Modulation

Before the Digital Audio Stationary Head (DASH) standard was accepted internationally for longitudinal recorders, Soundstream Corporation demonstrated a digital audio system using a pulse-code modulator (PCM) that fed a helical scan VTR. The recorder, contained in two small cases, was markedly less expensive than its larger counterparts because it used a 3/4-inch VTR as the transport and standard videocassettes as stock. The PCM audio information was recorded on the video tracks as a digital pulse code that was contained in an FM envelope. It left the two normal audio tracks free for code and reference information.

The video PCM method is still widely used today. The PCM module has even been scaled down and adapted for 1/2-inch Betamax (the F1 system), with a proportionate price adjustment.

Reproducing High-Fidelity Audio

The compact digital disc (CD) is widely acknowledged as the best way to reproduce high-fidelity audio—both in the home and for professional applications. Its ancestor, the optical laser disc, was developed expressly for use as a mass-storage and distribution medium for television programs.

The Sony Corporation invented a hi-fi format for its Betamax recorders with better specifications than most analog studio ATRs. It used the inexpensive helical tape transport, but retained the ability to record video simultaneously. The Beta Hi-Fi format records stereo audio as analog FM tracks in helical scans next to the video FM tracks (see Figure 7.1). The audio, while not digital, is so close to digital quality that even professional engineers are enthusiastic in their praise.

Of course, the makers of the VHS consumer format were quick to come out with their own version, which is of comparable quality.

Because of difficulties related to editing the adjacent analog FM tracks without inter-track crosstalk distortion, professional analog VTRs using an analog FM audio track are not yet available. It is indeed ironic that a $600 Betamax VCR can outperform a $25,000 1-inch VTR in a listening test.

AUDIO FOR FILM

As soon as computer-assisted editing became routine in television post-production, people in the film industry recognized the advantages of adopting the same techniques.

The venerable double-system sound method of film, traditionally most resistant to change, began to bend in the winds of progress.

Figure 7.1: Beta Hi-Fi Format

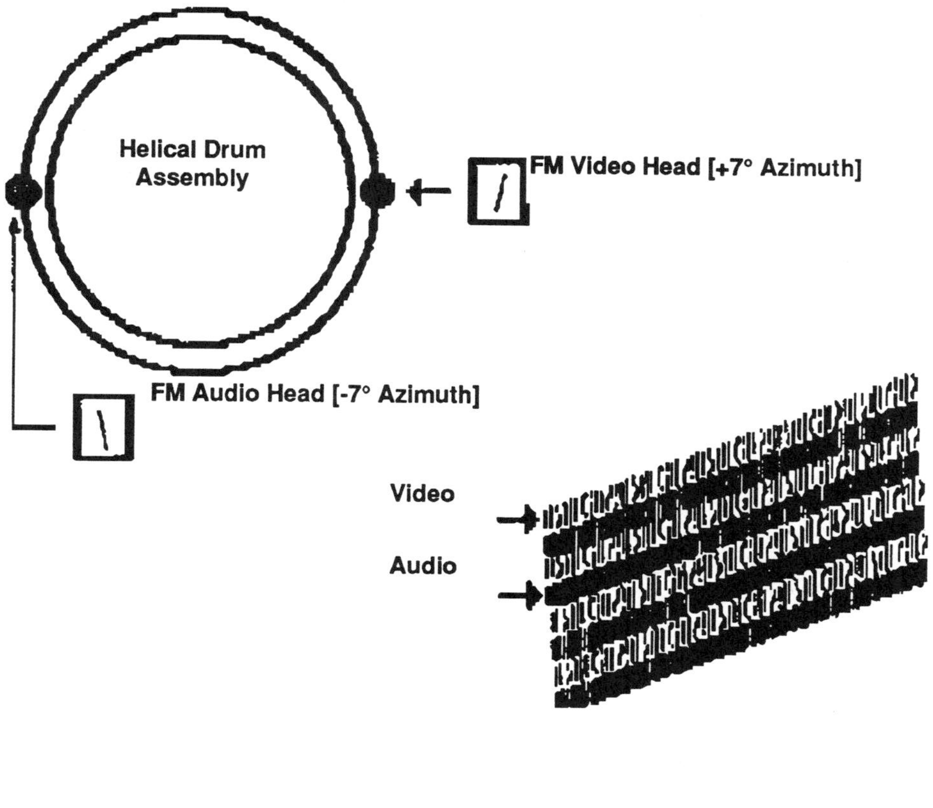

In 1980, a pair of young machinists doing business at Coherent Communications painstakingly rebuilt the shutter of a 16mm Arriflex ACL camera. The goal was to fit an array of LEDs (light emitting diodes) on the gate so a tiny portion of each frame's edge would be exposed outside the picture area, with a pattern. The optically recorded pattern, if readable, could represent time code.

Since time code is an audio frequency signal, recording it on a location ATR is merely a matter of replacing Pilotone information with code on the track already allocated as sync.

The only obstacle—the recording of code on film—was overcome by the Coherent team. Once they succeeded in retrofitting an LED matrix on the tiny aperture of a 16mm camera, mounting the optical encoder on 35mm and larger cameras was a simple matter of scaling up the housing.

The way the film hardware was adapted for code was probably the only way to encourage the film professionals to try it. Always reluctant to support such an "unholy marriage" of technologies, the film community had already been disappointed by previous attempts. Since the Coherent retrofit didn't commit the system irreversibly to code, there was minimum risk in testing it. The field operation of the hardware was unaffected; system controls and behavior during the shoot were the same for Pilotone or code.

Code *was* different during the editorial process, and *there* it was vastly superior. With code, *every* frame had a unique address that corresponded to a precisely defined relationship of the picture and track. For the first time, the syncing of dailies could be automatic. The clapstick was unnecessary. The efficiency of traditional film splicing was immediately increased.

Recently, more ambitious efforts at combining tape and film post-production have emerged, demonstrating further benefits. The reduced cost of computer-based editing systems and smaller, more accurate videotape formats have made the video editing room almost as commonplace on film lots as the Moviola. The ability of computers to accurately cross-index a video edit list to a negative-cutter's list makes a film rough cut virtually indistinguishable from a video offline master. "Cutting" a film on video eliminates the usual need for many duplicate film work prints, increases the amount of edits possible each hour an editor works and saves an appreciable amount of expense for the producer. By using video, dissolves and other transitional effects can be included in a work print for approval without the costly delays of the usual optical printing.

In fact, if a film is shot for television, there is no longer any reason *not* to post it entirely on videotape. Indeed, the networks no longer accept film release prints for regular series; they now insist the shows be delivered on tape.

A "yes vote" of major impact was expressed by the Arri Corporation when it bought the exclusive marketing and manufacturing rights to the Coherent film time code system. Other decisive votes have recently been cast by Eastman Kodak with its DataKode film (magnetic imprinting of SMPTE code on stock) and by Panavision Corporation. The new Platinum Panaflex series camera by Panavision has been completely redesigned for electronic post-production. It is equipped with onboard code and character generators, a TV-compatible digital shutter and a reflex viewing system that allows optical and electrical connection to a VCR for "instant video dailies." Film and video have been "married," despite any objections.

PROGRESS AND COMMUNICATION

Because of audio's importance to all of the communicative arts, it has been a common meeting ground for many different people working in widely varied specialties.

For example, the MIDI (Musical Instrument Digital Interface) standard for computer/musical instrument connection involved an international panel of musicians, computer designers, software engineers, instrument makers, audio equipment manufacturers and broadcast professionals.

The result in the case of MIDI is the interconnection and useful communication of equipment from various countries and manufacturers. For example, using the MIDI standard, an American-made computer, a Japanese synthesizer and a keyboard manufactured in England, all can be interconnected using a single inexpensive cable. The synthesizer receives data from the keyboard "telling" it which key is being played, and how hard. As the synthesizer responds to the keyboard input by producing sound, it sends the keyboard MIDI signal along to the computer while simultaneously adding its own patching and control-setting data. The computer reads the combined data stream and configures it for editing, automated sequencing, storage and printout in musical notation. The entire setup can be stored on a microfloppy disc and recalled for future execution from the computer terminal, which would then patch the synthesizer and play the song on the keyboard. The power of this type of automation is that each instrument "knows" how to "talk" to the other parts of the system. Manufacturers are now adding the accuracy of SMPTE code to their MIDI instruments—precision *and* power combined.

We are now the fortunate witnesses to a flurry of progress in the communicative arts and sciences. This progress and the achievements that come with it are possible when people recognize their common interests and participate freely in the development of new ideas.

A modern renaissance is underway. These are exciting times for media-conscious people. With the favorable climate established for meaningful dialogue among all the professionals within the broad category of communications, we can expect spectacular future developments. More exciting still, because of the increasing availability of new technology to all of us, we will be participating directly in the adventure.

CONCLUSION

Today, one of this country's highest-priority goals is the development of a standard for a high definition television (HDTV) transmission system. The pursuit of this goal is a matter of national concern, involving major corporations and top levels of the government. The international competition (especially from the Japanese) for dominance in HDTV has been compared to the race to the moon in such esteemed periodicals as Scientific American.

This race for technical innovation is reminiscent of the early pre-NTSC days. The same feeling of urgency seems to be enveloping the globe. But, today's world is very different from that of 1939. It's a lot smaller. We now know we are members of a global village, and have common needs which far outweigh our petty differences.

Today, television broadcasting is a rather mundane and commonplace part of our lives. The new vanguard of experiment and experience is the computer. In Chapter 8 we will explore the amazingly simple theories that may take us into a digital future.

8 Digital Designs

In Chapter 7, we referred to developments in television engineering that were closely related to digital technology. The invention and use of digital computers have touched every aspect of our lives, and where else but in television and other technologically intensive arts and sciences would one look for evidence of its impact?

Television was the first high-speed communications medium and is still the most widely used today. But despite this fact, television's mechanical design has remained virtually unchanged—the technology of the 1930s has served us much longer than anyone expected it would. But, before we plunge headlong into a look at the digital future, let us remember a poignant and humbling fact: almost every computer in operation today, from the common Commodore 64 to the awesome Connection Machine (developed at the M.I.T. Media Lab), communicates with the outside world through a standard, "garden variety" kinescope CRT.

SIMPLE MATHEMATICS

It has been said, all too often, that a computer accomplishes its functions by performing mathematical operations. While that truth is undeniable, it does not go far in explaining just how simple the theory really is.

The power of computers lies not so much in the complexity of the mathematics used as in the incredible speed with which the operations are carried out. To accomplish a wide variety of operations at an accelerated rate, the operations must be simple on the individual scale—the simpler the function, the swifter it may be accomplished. The starting point in this respect is the mathematical foundation on which the operations are built—simple mathematics.

The Base Ten System

For our discussion, it will be useful to forget everything you may have been taught in school math classes. Attempt to start here with a clean slate, throwing away the assumptions of the past.

The first thing to throw away is the assumption that numbering systems are devised merely to count things. A numbering system can represent more than a count. Counting is an arithmetic operation—addition, the accumulation of a sum or total. Most people learn the decimal, or base ten numbering system, which is natural when fingers do the counting. The fingers serve as symbolic reminders of the total. The use of symbols to represent numerical values is mathematics—math is a language. Higher mathematics come into play when a person uses ten fingers to symbolize more than ten numerical values—the symbols take on values beyond the intrinsic numbers obtainable through a simple count.

For instance, when a person counts 11 dollars, he/she runs out of fingers before the count is finished. The counter must devise a way to symbolize one more than ten, using only the ten different markers available.

In the base ten system, a digit may have one of ten possible values: 0,1,2,3,4,5,6,7,8,9. There is no single digit to represent ten. The value ten is an overflow—two digits. An overflow is a value that exceeds that of the highest single digit in the system. The value ten is, in fact, a notation meaning: "one overflow plus no single-digit remainder," written "10." Eleven would be an overflow plus one, written: "11." In these two-digit notations, the leftmost digit, also called the most significant digit, is a symbol representing the count of overflows; the rightmost, or least significant digit, represents the count up to or after overflows. When two digits overflow a third is added to the left to keep track of the second-digit overflows, and a fourth is added when the third overflows, and so on.

Most familiar numbering systems are configured this way, regardless of the base. There is always an integer (single digit) count, defining system capacity, with optional overflow counts to the left that are used to keep track of the number of times the integers have reset or cycled through all the possible base values.

LOGICAL MATH

The first computing machines were mechanical calculators used to manipulate normal base ten values. These machines were assemblies of wheels and gears, used to duplicate mechanical functions. The early automatic looms, which stored patterns for finished fabrics in the form of holes punched on wooden slats is one example. A mechanical model is termed an analog. The first machines of this type were analog computers.

Later electronic computers emerged including the Eniac machine, which used vacuum tubes. The tubes, and later transistors, kept track of values by acting as switches. Since a switch has two states, on or off, a two-value numbering system is desirable for notating its status.

The base two or binary system fills the bill nicely. It has two values: 0 (off) and 1 (on). In binary, the first overflow occurs at the value two, which is written "10." At a count of four, the second overflow occurs, written "100." In base two, each more significant digit represents a value twice that of the digit to its immediate right. Binary notation appears cumbersome because it needs more overflow columns to express a value than a decimal notation. However, it is inherently simpler within each column, and therefore, can be manipulated much faster by a machine composed of switches.

BOOLEAN ALGEBRA

The *binary* (two-value) *digits*, or *bits*, may represent much more than arithmetic results within the context of computer data. It can be useful to think of a computer as a pyramid of switches, with each switch's state (0 or 1) contributing to a larger configuration. The dimensions and overall characteristics of the entire switch array may represent almost anything the imagination could create.

Bits can be arranged in larger groups, called *bytes*, which can form words grouped into commands, arguments, variables, strings or expressions. All of these, though entirely constructed from the two numbers 0 and 1, are symbolic of much more. If the terms sound more like language than math, you're grasping the real significance. Algebra deals more with sentence (equation) structure and grammar than with numerical operations.

Part of this language is a system of logic, a specification of conditions or truths—the definitions of possible relationships between elements and values. The values 0 and 1 can mean "low and high," "false and true," "off and on" or "no and yes." The preferred logical system used for digital computation is Boolean algebra. It defines logical principles like those mentioned above that are especially well-suited to transistor-to-transistor logic (TTL).

In essence, Boolean logic systems are tailored specifically to be executed by electronic means—fast switches.

To compute values and relationships with simple switches, two binary digits are manipulated in various configurations. This is digital computing, or the processing of information in the form of digits.

Numerical systems built on the binary foundation are merely groups of bits recognized as individual chunks of data: the octal numbering system uses bytes of eight bits' length; the hexadecimal system (6 + 10) uses 16 bits; and the binary coded decimal (BCD) system uses four bits to display a decimal.

THE MICROPROCESSOR REVOLUTION

Growth in the electronic computer industry was never slow, but the cost and bulk of early designs kept the computer confined to the laboratories of government and corporate institutions.

When Bell Laboratories demonstrated a working solid state circuit, the transistor, the first gigantic step toward cost and size reduction was realized. Hot, fragile, expensive and bulky vacuum tubes could be replaced with components that worked more reliably, put out better signal quality, required less power and occupied a fraction of the physical space. Moreover, because the new transistor was made almost entirely of essentially inert crystalline mineral compounds, it cost very little to mass produce.

Bell Labs had the resources to work on several large-scale development projects simultaneously. Almost as soon as the first transistor prototypes appeared, work started on integrated circuits, in which many discrete transistors and their associated wiring were miniaturized and combined on a single chassis of drastically reduced size.

The success of Bell stimulated the growth of a number of smaller companies and the interest of many large corporations along similar lines. The growth of the computer industry skyrocketed. The massive miniaturization afforded by today's very large scale integration (VLSI) techniques has put the personal computer in the home.

Computers with many times the speed and power of the first large prototypes are now contained on a single one-inch chip of silicon compound.

COMPUTER ARCHITECTURE TODAY

The internal architecture of today's computers starts with the *central processing unit* (CPU), which acts as the data processor (see Figure 8.1). This processor is composed of an arithmetic logic unit (ALU), or "brain," which performs all the arithmetic and logical operations. The ALU runs the coded program instructions for executing operations. Often, the program code is stored on another chip that is either built into the ALU or stored adjacent to it. This type of integrated circuit (IC) is called a read-only memory (ROM) since once loaded, the data stored on it is not erased. Permutations of basic ROMs evolved as the need to modify or change operating programs developed. These include programmable read-only memories (PROMs) and erasable programmable read-only memories (EPROMs).

Within the computer there is usually some provision for storage of data undergoing operations, or "scratch pads" the processor may use during its work. This is the memory of the machine. *Random access memory* (RAM) ICs store this type of data, as well as different programs the system may call for. These ICs may be written into and read from as needed. Each bit has a specific address where it is stored and from which it may be recalled.

Figure 8.1: Microprocessor Design

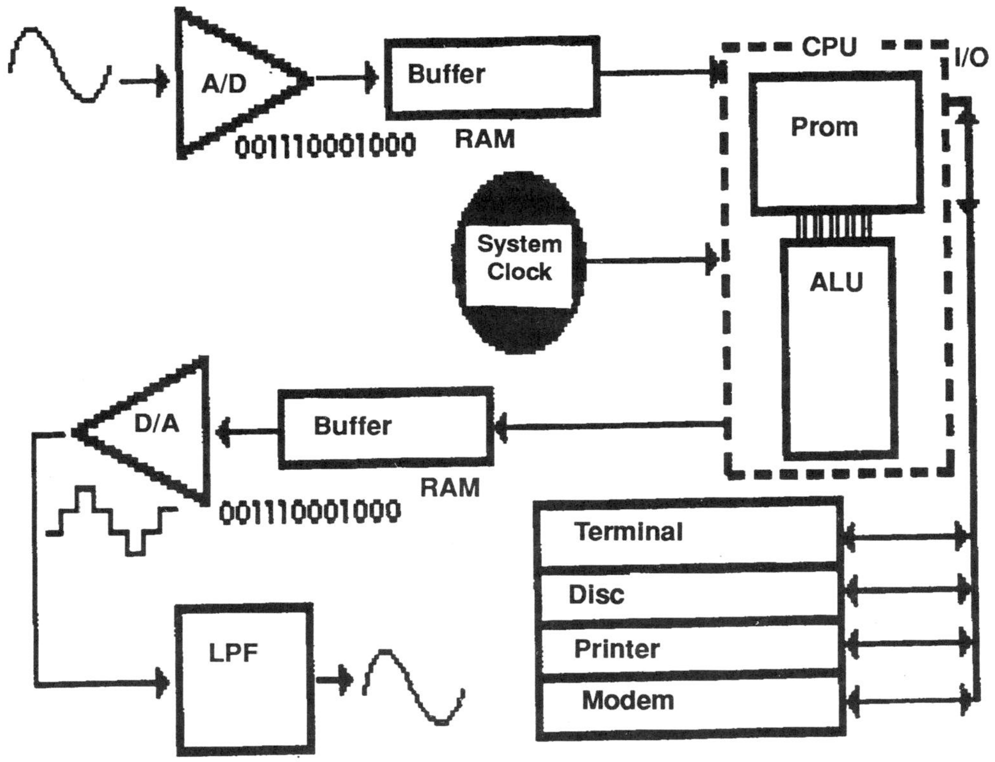

The computer system also contains an input/output (I/O) bus, the communication line through which the machine is connected to the outside world. The computer uses this port to accept input from devices such as a terminal keyboard and output information to devices such as the terminal CRT. Disc drives, printers, modems and other communications devices are also connected to the computer through the I/O bus.

Overseeing all aspects of computer operations is a system clock, which determines when the different parts of the system are free to execute new instructions. As data is moved through the system and as operations are performed, the clock acts as a "traffic cop" to prevent logjams. The frequency of the clock provides the speed limit for the system. For example, the computer used by the author to write this chapter has a Motorola MC68000 processor with a clock running at 8 MHz. It can process eight 32-bit numbers every millionth of a second.

ANALOG TO DIGITAL

Up to now, we have been studying signals in the analog domain. An analog is, literally, a mechanical copy of nature. Analog watches are copies of sundials; the hands represent the shadows cast by the sun. A digital watch does not move a set of hands around a dial. Instead, it displays numerical digits that represent time values computed by counting the vibrations of a crystal oscillator.

Digital signals are numerical values. Because storage of numbers is less demanding than storage of a waveform analog, there are certain advantages (and disadvantages) to conversion of analog signals into digital values for storage.

The first widespread use of microprocessors in broadcasting was for audio effects processing. Because the digital audio devices were going to be operated in an analog environment, the acceptance of this technology hinged on ease of installation and use. The first devices were "black boxes" with computers at their cores. These are the dedicated digitals—computers "disguised" behind user-friendly familiar control panels.

A useful example is the Eventide 949 Harmonizer. The 949 is actually a very versatile audio digital delay line under the control of a microprocessor running a very clever program.

DIGITIZING AUDIO

The 949, like any other computer, does its magic by processing numbers. To get numbers for processing, the 949 must convert analog audio into digital data. The circuit for this purpose is the analog-to-digital (A/D) converter. Each pulse from the system clock triggers the A/D to sample the audio signal at the input. The sampling operation consists of measurement of the input signal's amplitude at that instant, and assignment of the measured value to a digital quantity, a binary number. The number is the digital equivalent of the volume of sound at that precise moment.

The dynamic range of the system (the difference between the strongest and weakest signal output) is determined by the range of digital values that can be processed during each sampling period. The determining factor here is the number of bits available for value assignment—the longer the word, the finer the distinction between one amplitude and the next. For example, an eight-bit processor can process no more than eight bits during each sample. The number 00000000 represents silence, and the loudest sound would be assigned a byte of 11111111. In between these extremes lie 256 different possible volumes.

Because A/D sampling is basically a one-dimensional operation (amplitude), the frequency response of the digital system is dependent on the number of samples taking place each second. In A/D conversion, a rule called the Nyquist Criterion dictates the minimum clock rate as two times that of the highest input frequency to be sampled. In the 949 Harmonizer, the desired output bandwidth was 20 kHz, so the clock had to operate at a rate of at least 40 kHz.

The 949 was designed to output signals at 1000 different possible amplitudes, so the digital sample had to be ten bits long.

The input signal emerges from the A/D as ten bits of digital data at each clock pulse of the system. Once the signal is turned into data, it can be acted upon by the processor. Since the audio is now only a number, anything that can be done with a number can be done to the audio—with some startling results.

The RAM Buffer

After A/D conversion, the data goes to a RAM array for short-term storage. In the 949, 11 16-kilobyte chips allow .4 seconds of audio data to be held in storage. Any or all of the data can be loaded or unloaded from this array during each sampling period. Output of the data on a first-in/first-out (FIFO) basis results in a 400-millisecond time delay—in audio terms, a pronounced echo. The amount of RAM used during a sampling period is selectable, allowing shorter delays to be programmed. The simple digital delay line (DDL) function is the least spectacular of the effects possible with the 949.

In real-time systems like the 949 Harmonizer, RAM is always used as a buffer, a "holding area" for data waiting to get to the processor. The buffer memory may also be used in a variety of other ways. However, its main function—avoiding traffic jams on the way to the ALU—prevents loss of data "in line" to the output of the device. Manipulating the RAM stack data without any arithmetical functions can be exotic: if data is read (output) from RAM in reverse of the write (input) order, the output audio is time-reversed, the digital version of playing a tape backwards.

The Central Processing Unit

The CPU is where the calculating heart of the computer, the ALU, resides, along with the instructions contained in the operating program. By subjecting data to arithmetic operations, the output can be changed in seemingly magical ways.

For instance, if an announcer's recorded voice is too high-pitched, the analog solution would be to play back the tape at a slower speed. Of course, playing at a slower speed not only lowers the frequencies of the signals, but makes the announcement longer in total duration.

Again, the 949 Harmonizer can adjust the pitch *without* lengthening the total duration. It down-pitches digitally by reading the data at a lower rate than that used to write into storage. The total duration of the signal is preserved by digitally splicing the pitch-modified segments of data during the silent pauses and deleting any silent samples. This operation is a frequency-division of the signal with subtraction of silent samples. The 949 can down-pitch a signal by as much as two full octaves this way. It can also up-pitch a signal as much as one octave by frequency multiplication and sample-additive splicing (the inverse of the former arithmetic).

This sounds very complex, but it is really just a great number of *simple* operations taking place at very rapid rates. For example, multiplication is nothing more than repeated additions; and divisions are nothing more than repeated subtractions. During any one sampling period, the computer is processing a ten-bit number made of ones and zeros. In the 949, the system clock runs at 88 kHz. The magic is made possible by virtue of the ALU's ability to do each simple operation in 1/88,000 of a second.

Real-time dedicated digital processors such as the 949 work this way. In the video domain, the Ampex ADO (Ampex Digital Optics) processes video bandwidth signals in basically the same way. The video signal range of 4.2 MHz requires a much faster clock and a bigger processor, but the computations and theory are virtually identical to those used in audio. Once a signal is *digitized*, it is nothing more than a binary number, and operations performed on it have the same effects regardless of whether the value represents audio or video.

DIGITAL-TO-ANALOG CONVERSION

The desired output is a modified analog signal. After it has been processed, the digital word representing the new signal is routed through a buffer (traffic control of time-modified data) to a digital-to-analog (D/A) converter. The digital word is output from the D/A as an analog voltage appearing at each clock pulse. This voltage is the output audio (or video) signal. If the conversion to and from the digital domain is reasonably transparent, the signal emerges from the computer with little or no distortion. Any artifacts that are added by digital processing should be intentional—all other signal parameters should be unaffected by the process. Unsurprisingly, zero-error systems are just as impossible in digital designs as in analog.

DIGITAL DISTORTION COMPONENTS

Consider the difference between a wristwatch of analog style and a digital watch. By virtue of the analog's use of hands moving around the face, many time values are indicated without being named. The set of possible time indications is nearly infinite, limited only by the thickness of the hands' points, since most of their travel occurs over places on the watch face that are not imprinted with numbers. Indeed, most analog wristwatches indicate "in-between" times as mere hash marks.

Digital watches, on the other hand, must display *all times* as discrete numerical values—there are no analog hands to indicate values by positions between printed times. If a digital watch is to split a second into tenths, it must display each tenth division. Thus, the set of possible time indications for a digital watch is severely restricted; it is limited by its number of digits. A tenth-second digital clock cannot give any indication of 1/100 second fractions. In some applications of digital systems, this non-continuous nature of data representation may become glaringly apparent, especially when continuous-wave signals are being processed.

A sine wave, for instance, is a continually varying waveform with an average value that is derived from an infinite set of points. Theoretically it is a perfectly smooth curve. Mathematically perfect sine waves do not exist. Nonetheless, the approximations of sine waves and other curved signals that occur in nature represent the most difficult functions for a digital system to handle.

The digital segmenting of curve functions occurs in both wave dimensions: the sampling clock breaks the curve along the vertical axis, and the quantities assigned during a sample break the curve horizontally. (See Figure 8.2.)

Because only defined quantities are digitized, a value falling above or below recognized values is assigned the nearest quantity. The result is a sine wave approximation at the output with a stair-step component superimposed on it.

The values are assigned only during a sampling period. Therefore, significant portions of the sine curve, which is changing continuously, are omitted from the data stream. This produces a similar stair-step component on the perpendicular axis of the output. This stair step is the frequency of the sampling clock.

Anti-aliasing

The distortion of amplitude and wavelength values are unavoidable and unique to the digitizing process. However, they can be masked.

Quantizing Error

The distortion of amplitude values is handled by adding more bits to the word to

Figure 8.2: A/D Sampling

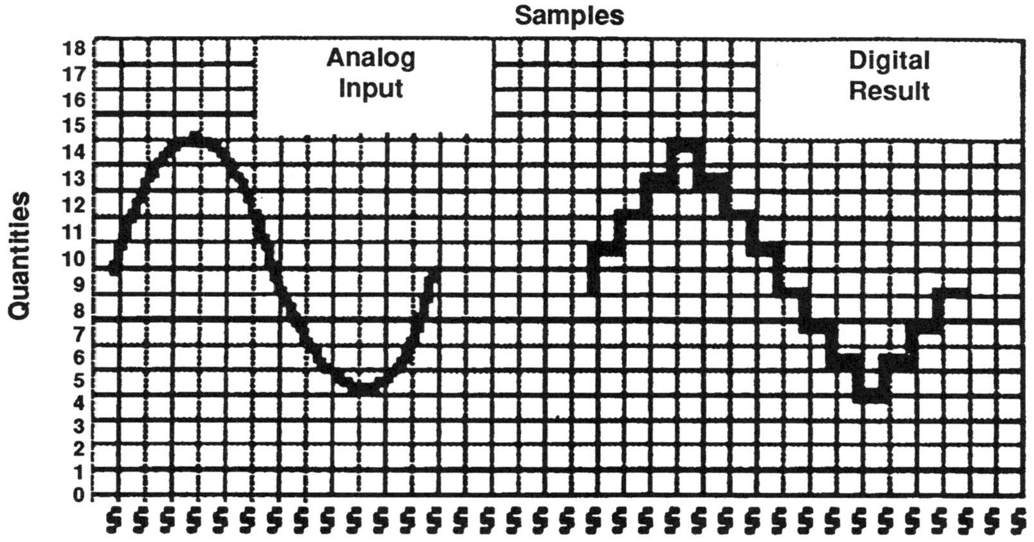

break it into smaller segments—each LSB (less significant bit) added reduces the error by half. Also effective for some applications is average-value assignment, floating-point quantizing and other mathematical processes for "smoothing" curves digitally.

Aliasing

The distortion of wavelengths by the system sampling clock is handled by using a low-pass filter after D/A conversion to eliminate the clock pulses from the output. Wavelength distortion also benefits from specially designed mathematical techniques, and it is now customary to "lump" corrective "smoothing" measures for both amplitude and wavelength errors into a single terminology: ***anti-aliasing***.

Improvements to Digital Hardware and Software

As the speed and efficiency of digital processing continues its inevitable improvement, the effect of errors will diminish proportionately.

Hardware is not the only variable in this equation, however. Software, the programmed operating instructions for the processor, is every bit as important. As of this writing, the possibilities offered by revolutionary equipment design have far outdistanced the software available to make their operation useful. At this time, we await insight and inspiration on how to get our incredible machines to behave with at least some hint of the diversity, flexibility and intelligence of the human mind after which they were modeled.

Current real-time digital video processors include Grass Valley-NEC's Digital Video Effects (DVE), the Ampex Digital Optics (ADO), Quantel's Mirage and Vital Industry's SqeeZoom. There are many similar lower-end manipulators, all based on designs centered around a digital TBC/framestore that is controlled by one or more microprocessors. Each of these machines runs programs to interpolate the positions and brightness of processed pixels within the two dimensions of the CRT corresponding to the three-dimensional images they represent.

Other computers used in production are more conventional. The computer-assisted editing systems, from the glamorous CMX-6000 to the humble Convergence ECS-90, are more like business machines than magic wands. These machines process edit decision lists and synchronize video recorders, tasks not very different from the monitoring of your account balance at the local bank.

New digital machines such as the Bosch FGS-4000 and the Quantel Paintbox, though not real-time, are adding new dimensions to electronic production by allowing video to be treated like paints and inks. With these machines, video artists can literally draw on the CRT as if it were a blank canvas, using palettes with as many as 16 million different shades, colors and textures.

Digital videotape recorders (DVTRs) are emerging, capable of recordings that are more faithful than anything before possible, and impervious to the loss of quality always associated with copying analog recordings. In addition to the BVH-2000PCM from Sony and the VPR-3 from Ampex, the Quantel Harry and the Abekus A60 and A62 are noteworthy in that they use non-traditional magnetic storage media: computer disc stacks.

Because the recorded signal of a DVTR is binary data, dubs (duplicates) of the tape do not accumulate noise as quickly as analog dubs. DVTR users commonly dub 20 or more generations down from the original with no detectable degradation.

The two accepted standards for digital video recording are the D1 and D2 formats. Both convert video into a binary data stream and record it on a cassette. In D1, the video is input and output as RGB signals, and is encoded for broadcast downstream. It allows flexibility and preserves signal quality, but distributing the three analog output signals requires three times the wiring of a normal encoded NTSC distribution system. In D2, the recorder uses NTSC or PAL composite signals, and is ready to record and play through standard wiring.

Digital audio recorders from a wide variety of manufacturers are available in two standard formats. The Digital Audio Stationary Head (DASH) format is the oldest, about 10 years. DASH machines are commonly found in recording studios, usually as multitracks from two channels to 32 channels. They look like analog machines, but turn 1-inch reels.

The other format, Digital Audio Tape (DAT), is becoming popular in smaller machines for portable field use. DAT will probably be adapted into consumer machines. A controversy has recently developed concerning the DAT machines. Record producers are worried about piracy and bootlegging. If any DAT copy is as good as the original master, why buy a tape if you can just dub it? The big record labels have tried to embargo trade against DAT for consumer products. The debate is still active in Congress. Meanwhile, the producers are trying to come up with a copyguard signal to scramble the DAT data and make it unsuitable for unauthorized duplication.

The promise of high-density storage using optical media, such as laserdiscs and compact discs, will further improve and refine the digital recording process. The use of optical rather than magnetic media will surely revolutionize the transports used in recording, making recorders less mechanical and prone to wear and tear, and eliminating many of the large moving parts that produce errors and distortions today.

THE REVOLUTION IN POST-PRODUCTION DESIGN

The recent development of all-digital recording systems signals a monumental change in the modus operandi of production and post-production of both theatrical features and broadcast television programs. Physically, the post-production process is the integration of many different location elements into a single film or tape—the edited show. A significant amount of the time and effort is expended in accessing and reaccessing the material. Reels are loaded, used and unloaded many times throughout the process. If a reel is loaded only once for all editing, the process is much more efficient. The B Mode or checkerboard sorting of a list for auto-assembly is a step towards reducing wasteful reloading. (See Chapter 6: Videotape Editing.)

Although flexibility has improved dramatically, the efficiency of the process has only been marginally enhanced—even the CMX 3600 does little more than mimic the physical splicing process by controlling 10 machines instead of one or two. And, adding machines to a session increases costs. These "octopuses" are doing more of what has always been done—better than it has ever been done before, but *only* that.

The Fairlight Corporation in Australia and E-Mu Systems in Santa Cruz, CA, have designed digital audio systems that process the program signals directly without the normally prerequisite tape machines. The Fairlight Computer Musical Instrument ($70,000) and the E-Mu Emulator ($13,000) capture samples of actual program material in RAM and subject them to digital processing directly, storing the results on common floppy discs for recall and further processing. This powerful combination of processor and digital recorder had an immediate impact on the film, video and record industries. Other manufacturers, like New England Digital (Synclavier) immediately followed suit with their own all-digital designs.

The most revolutionary current audio design was the SoundDroid from LucasFilms' DroidWorks. The proposed idea was to replace the tape transports, mixing console and all peripherals (such as equalizers, compressors, limiters, echo chambers, DDLs, even musical instruments) with a single system of processors and high-density storage devices. In other words, all the elements of a show would be directly loaded into the SoundDroid, out of which a finished, fully polished product would emerge.

This system would have been operated from a computer terminal workstation with "soft" controls, instantly reconfigurable to suit any operator or task. Because all program elements and the "devices" for program manipulation were digital data, the editorial possibilities would be limited only by the cleverness of the software and the system's storage capacity. In the 'Droid, the physical liabilities of today's methods simply did not exist. Unfortunately, however, the SoundDroid suffered the fate of the other products of LucasFilms' DroidWorks: abandonment. Although 'Droids will be developed and used for LucasFilms projects, no commercial licensing or manufacturer is planned as of 1989. Fortunately, the basic premise of the SoundDroid, direct digital manipulation of the program, lives on with other manufacturers' prototypes.

Of course, these sweeping changes in the narrow applications associated with the production and post-production of television and film are mere reflections of broader technological improvements in the materials and designs of the next generations of digital communications and computational devices.

FROM ELECTRON TO PHOTON

The projected theoretical limit of current silicon-based circuit technology is estimated to be only a decade from practical realization. As of this writing, the speed and capacity of digital communication and computing is in the gigahertz/gigabyte range (billions of operations). Before exhausting the possibilities of silicon, a hundred-fold increase is possible, but scientists are already working on other technologies, such as gallium-arsenide chips, VLSI semiconductor lasers and photonic memory chips, all of which promise thousand-fold increases immediately.

The developing hybrid technologies have one major goal as the key to future improvement—the maximum first-level (on the chip) interconnection of component circuits. It is here that the cost of connection is lowest, and the efficiency (thus, speed) of the resultant higher-level circuits is greatest.

All solid-state circuits used in communications today move electrons through wires. Electrons are particles with measurable mass (hence, inertia), size and a set of pronounced electromagnetic properties. We are approaching the point where many of these characteristics, which have been very useful so far, are now beginning to get in the way of further progress.

Of great interest to the communications industry is the replacement of existing copper and aluminum networks with higher capacity transmission lines. Undersea cables currently crossing the Atlantic Ocean are capable of carrying 10,000 conversations per line. These cables use repeater stations at regular intervals (about 25 kilometers) to regenerate the electromagnetic pulses, which are greatly attenuated by the wire.

If the transAtlantic cable were replaced with one fabricated from optical fibers carrying laser light pulses instead of electrons, the capacity would be increased four-fold with the use of a thinner cable that would run distances of 65 kilometers between repeaters. Because the signals transmitted through an optical cable are photonic, they are insensitive to magnetic interference and noise, immune to electrical interference and virtually incapable of short circuits, catastrophic overloads or internal heat damage.

Signal systems based on photons as the transmission particle, instead of electrons, will enjoy incredible advantages, but the switching, amplifying and controlling hardware for such systems is in a very primitive stage of development. When we have built photonic systems capable of using light the way we currently control electrons, today's communications science will appear as crude as Bell's first telephone.

The interest in photonic technology has encouraged researchers to re-examine classical perceptions of nature, creating not only new machines and theories, but new concepts of science. More useful models of human sensory and brain functions are emerging. It appears likely that photonic computers will be radically different than what we've seen so far. More closely resembling the human brain, these optical-neural arrays will be able to recognize patterns, fill in missing data, correct errors and possibly learn on their own. New sciences like *chaos*, the study of hidden order in random events, unite computer scientists with biologists. New computer designs such as the Connection Machine are handling graphic computations impossible for a Cray XMP supercomputer at a fraction of the expected cost.

While we continue to strive towards perfection, it is indeed inspiring (and seductive) to contemplate the discoveries, inventions and methods that lead the way to the spectacular improvements we will design for the fuller enjoyment of our lives. We are driven to new achievement by our hunger to do things better and our need to understand ourselves and our universe more completely. Occasionally, our ardor in this pursuit can blind our objective vision, obscuring what seems to be the "first law" of all technological evolution: only those changes which are affordable can be realizable.

For instance, no knowledgeable engineer would argue with the assertion that there are systems superior to NTSC television in imaging quality. What is arguable is the feasibility of this particular method for economically realizing the practical application of visual communication for the masses. No improvement in the system can override the ultimate consideration of mass deployment, which is inversely proportionate to the cost.

No new idea, however eloquent or ingenious, makes it into the development lab without first having the potential for emerging as a working, practical product that can be marketed for a recoverable and reasonable cost.

THE PC REVOLUTION

As we await the arrival of the next generation of supercomputers, technology refuses to stand still. The cost of computing power is falling almost daily. Incredibly, the idealistic notion of sharing the benefits of this progress with the general public is actually happening. The personal computer (PC) is probably the biggest and fastest route for new developments entering the real world. It satisfies the critical requirement for cost effectiveness magnificently.

The price/power ratio of the PC is an incredible enticement for software developers. Applications programs for PCs now can execute tasks formerly reserved for mainframes and minicomputers. In some cases, today's PC can run faster and more accurately than its older "brothers."

In the past few years, broadcasters have seen the PC take over many areas of activity. Callaway Engineering (CE), for example, is marketing a videotape editing system with striking similarities to the CMX 340X. There are two major differences between the CMX and CE systems. The CE architecture uses an IBM XT personal computer instead of the DEC PDP-11 minicomputer used by CMX, and the price of the CE is less than half that of a CMX of the same configuration.

Character generators, paint systems and more recently some real-time effects systems have taken advantage of the PC, with the same cost benefits.

In fact, some larger-scale system designs have abandoned the traditional large single processor in favor of an array of multiple microprocessors—essentially, working a number of PCs in parallel. The most promising new supercomputer prototypes use this "team" approach to computing. When a group of processors shares the load, each working on a different part of the problem, the system spends less time idling. The result is a much faster solution. The Connection Machine is this type of parallel processing device. It uses 65,000 microprocessors, each with its own small RAM. Any single processor in the system can address any other processor for needed data—the first electronic computer that begins to truly resemble the physical wiring found in the human brain.

It is now easier than ever to put a PC at your fingertips. Because of the extensive use of the PC in business and in the development of new computer products, a PC in your home today, is a good investment in understanding tomorrow.

CONCLUSION

Since our opening discussion in Chapter 1 we have followed an incredible path in tracing the evolution of the TV signal. We've gone along for the ride as photons were converted to electricity, sent through miles of wire, encoded, modulated, filtered and amplified. We have watched the signals turn into electromagnetic pulses and fly through the air. We have recorded these signals on tape as RF envelopes and rearranged them through editing. We have switched them, synchronized them, time-base corrected them, time coded them and digitized them for processing by computers.

By staying on track with the signal we have encountered all the machines and processes used within the TV system. In Chapter 9: System Mechanics and Troubleshooting, we will examine a few of the oddities that exist within this complex TV environment and learn how to use our knowledge of the system's characteristics to solve everyday problems.

9 System Mechanics and Troubleshooting

A thorough grasp of the basic operating principles of a complex system like NTSC television is a prerequisite to solving the problems rountinely encountered by television professionals throughout their careers.

In television, especially broadcast television, the urgent need to get around obstacles quickly often supersedes the calm and thoughtful analysis required to get to the heart of most problems. As the system evolves and increases in complexity, its complications tend to stack up one on top of the other. In the end, these often subtle problems are traceable to inconsistencies designed into the original, simpler system. If TV engineers expect to understand the root causes of the problems they face, they must never lose sight of the basic structure of the system with which they work.

COLORFRAMING, SCH AND RS-170A

Horizontal Shift

A good example of a subtle problem which is rarely understood is the pesky horizontal shift (H-shift) at an edit. This picture disturbance usually occurs when you are attempting to insert a matched-frame dissolve (same point on playback and record VTR reels) into a previously continuous scene.

At the in-frame point, the picture jumps suddenly to one side. The jump, only a few thousandths of a line's length, is too sudden and tiny to be a camera move. Nonetheless, the effect on the viewer's eye (especially that of the director, sitting in the next seat, paying $400 per hour for post-production) is terribly distracting and telegraphs the effect just behind it.

You check your source tape; no shift in there. You redo the edit, trimming a frame on either or both VTRs in both directions, only to find the problem getting worse. Stranger still, you may redo the edit without adjustment, only to discover that the in-frame is perfect, but there's a shift at the out-edit.

The VTRs seem to be unable to duplicate their former lock relationship. Now the shift is married to the show. The only remedy at this point is to hide the shift by moving it where it will not be easily seen. This might involve a number of edits, each one bringing the shift closer to the nearest hiding place—a cut to a totally different picture. Sometimes, the shift is so pronounced that the *only* place to stash it is in black.

You examine the list carefully, finding no error in the time code addresses corresponding to the point. Your assistant editor checks VTR lockup and performance at the edit and confirms that all parameters are normal. Maintenance is called, and the engineer discovers no abnormalities, yet the picture is taking a noticeable and annoying "hit" at the edit.

When all component parts of a system are apparently working within expected tolerances, an error of this sort is usually found to be *systemic*, that is, a cumulative result of the overall configuration—a sum of the parts.

The problem appears to have no "handle," because the handle is hidden behind other aspects of the system that are not obviously related to it.

H-shifts were never a complaint before computer-assisted editing. The reason is that before SMPTE time code, editors did not attempt to make match-frame edits. A slight shift in an edit to a totally different scene is invisible. The shift is apparent only if no other element of the picture changes at the edit point, and these conditions are possible only in the frame-accurate environment of computer-controlled edit suites.

A Process of Elimination: Colorframing

Backtracking further in the examination of the H-shift problem, let's scrutinize time code itself. The basic nature of NTSC scanning and its relationship to code is that *both* fields at *each* VTR's in-frame match point have the same address. SMPTE code does not distinguish between odd and even fields; its count resolves frames as pairs of fields. If the shift were vertical instead of horizontal, this might be the end of the road, since an odd/even field discrepancy at the edit would produce a serious interlace disturbance resulting in a pronounced vertical jitter or a sudden reversal of color hues in the picture. However, the symptom could be quite subtle and almost invisible to the untrained eye as a tiny shift in the picture exactly one line in height. This ambiguity is called a *color-framing error*. It is a result of the 180° change of burst phase that occurs on each TV line because the 3.58-MHz color reference frequency represents an odd multiple of one-half line. (Refer to Chapter 3: NTSC Color.)

Phase reversal was purposely designed into the NTSC system to suppress transmission difficulties intrinsic to color broadcasts. When color phase is reversed each line, any erroneous 3.58-MHz component will cancel on adjacent lines to suppress display of the error. The chosen reference frequency also produces the same phase-reversal between frames. So, any line of the current frame will be color-reversed compared to the same line in the frames before and after it. While this toggle of color phase is desirable for *transmission* compatibility, it presents difficulties in editing because of the segmentation of the video signal. The inherent limitations of mechanical precision in traditional VTR transport designs are most apparent in color recording and editing, where it becomes clear that the analog VTR is capable only of monochrome framing accuracy. A VTR cannot change the timing of any single element within the composite signal it records. The TBC may shift the relationship of the entire composite recording to match house reference, but the *internal* timing of signals within the envelope as recorded cannot be changed.

When the edit is examined with a vectorscope (a precision waveform monitor that displays a global map of the signal's chrominance envelope), the colorframe error is visible as a sudden 180° swing in burst phase.

By using color vector detection during the pre-roll lockup sequence, one of the VTRs can be vertically reframed by "bumping" the capstan servo. There is a 50 percent chance that the capstan will re-lock to the opposite field, and therefore to the opposite color phase. Originally, editors had to manually hit a capstan speed override switch to force servos to re-lock until the correct burst vectors appeared on the 'scope. Fortunately, manufacturers of VTRs and computer editing systems automated the process to ensure predictable field alignment.

The alignment of line-to-line burst phase is accomplished at the expense of edit resolution. That is, out of 30 frames present during each second, only half are properly phased for stable color editing. VTR manufacturers added a 15-Hz colorframe identification pulse to the control track signal. This pulse is used during editing to field-lock the record and playback VTRs' servos to assure proper field parity in the edited signals. Computer-assisted editing systems use the colorframe ID pulse and the relationship of SMPTE addresses (odd/even or odd/odd) to maintain correct color edit phasing of source and master tapes to within 180°. The speed and accuracy of automation in handling this signal alignment chore is practically invisible in all but the most severe cases. In our particular case, we find no colorframe error—the fields are aligned. What we do discover, however, is a medium error in phase at the edit—exactly 90°. Though not enough of an error to qualify as a bad colorframe, this phase displacement is what is disturbing the picture.

Subcarrier-to-Horizontal Sync Timing

A color error of 90° is precisely halfway between good and bad. If corrected by a "bump" of 180°, it will remain a 90° error no matter how the servos re-lock.

The lack of any formal standardization from manufacturers or SMPTE on the absolute measurement of subcarrier-to-horizontal (SCH) sync timing makes color phase errors of 90° or less on videotape recordings commonplace. Since the VTR is limited to shifting the entire sync period, any timing discrepancy between signals internal to the sync interval becomes visible. When mismatches occur between the reproduced (off-tape) sync-to-burst timing and station sync-to-burst timing, the VTR can match only one of the parameters. If subcarrier phase is corrected, horizontal timing will suffer. If H-timing is matched, color will change. Since color shifts are considered more disturbing than slight picture movement, the TBC usually "settles" for the H-error. It is this characteristic of composite FM recording that produces the annoying H-shift error.

The SMPTE RS-170A Standard

The use of digital TBCs in modern analog VTRs has reduced the impact of the H-shift error by allowing the output video to be shifted horizontally relative to burst phase in increments as small as 35 nanoseconds. This time period corresponds to 45° of burst rotation. Lesser errors are undetectable under normal viewing conditions. In addition, SMPTE has published a recommended standard (RS-170A) subcarrier-to-horizontal (SCH) sync measurement for the composite signal. Although compliance to RS-170A is not mandatory, the industry recognizes the benefits of such a standard and is voluntarily implementing it in new equipment. Component (RGB) digital and analog recordings, of course, do not store the video and sync signals as a "married" composite, so they can effectively "side-step" the entire trap.

THE "SYSTEMS" APPROACH TO PROBLEM SOLVING

In order to sense and overcome the inevitable problems of television, a good theoretical understanding of its origins is essential. Many times, a "glitch" in a picture that is caused by a technical factor may be solved by a production solution more quickly than a "frontal attack" from engineering, and vice-versa. Although technology appears to simplify many of the aspects of television production and engineering, there is no substitute for a working knowledge of the basic mechanical nature of the system.

In television, technical and aesthetic aspects are so interdependent that there is no real distinction between the craft and the art. The most successful editors are those who are capable of poetic interpretation while maintaining technical quality. The best directors, though never seen with a screwdriver, can tell the difference between a cameraman's "bobble" and an SCH error.

SYSTEM CHARACTERISTICS

Other "glitches" in the NTSC system plague editors and engineers and require repeated explanations to bewildered clients. For instance, directors are often perplexed by the jagged edges and noise that may accompany alphanumeric characters in titles.

Figure 9.1: NTSC Scan Curve Distortion

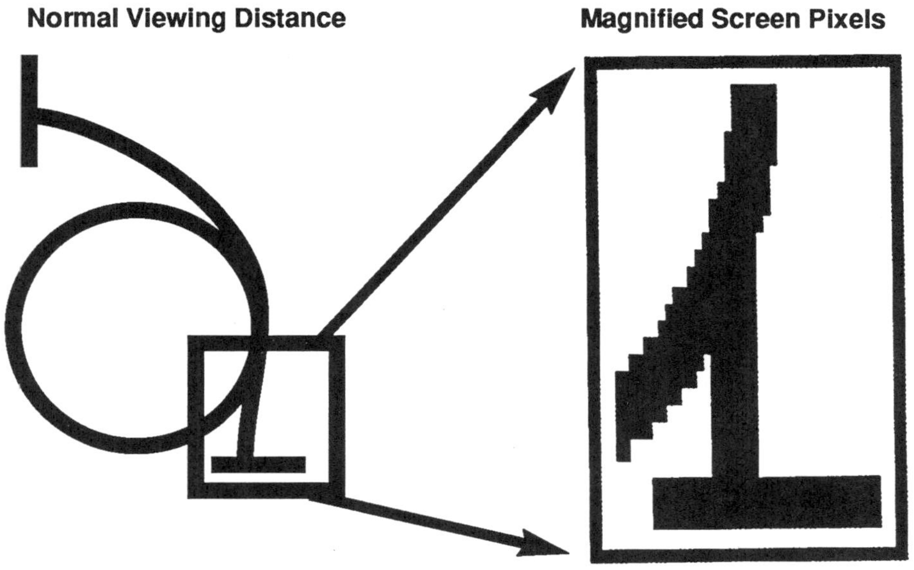

NTSC Scan Curve Distortion

The reason for the jagged edges becomes clear when you examine how the raster is scanned in purely mechanical terms. Using Figure 9.1 as our example, you can see that since the scanning beam is moving horizontally at a very rapid speed, curves appearing on the screen are actually a series of separate dots, stacked at positions offset from the top/left of the display. The beam *never* draws anything but a dot or a straight line, so any displayed curves are the result of stacks of many lines—an optical illusion.

Because the scanning dot is not infinitely small, as a theoretically perfect point is, it has measurable width. This width produces an error not unlike a digital quantizing error. The smoothness of any curve can be approximated only within the measurable limits of the technology. Any linear scanning system will fall short of producing perfect curves.

Noise appearing with text is invariably seen to "crawl" diagonally, or sometimes vertically, through the characters. If examined carefully, the frequency of the "crawl" is always found to be a harmonic of 3.58 MHz. If you are saying to yourself, "Aha! The noise must be chroma!" you are absolutely correct. Filter the color information out of the characters and the noise disappears, unless the characters are so small that they produce false 3.58 components when scanned.

There are simple solutions to these complaints. Reduce the chroma to limit the noise and use larger text to reduce the linear scanning error. As a professional, you are expected to have answers to your clients' simple questions. The answers may not be what the client expects, or wants to hear, but they *should* be the correct answers. You may discover a problem using a digital character generator manufactured in 1987, but its solution may be hiding back in the mechanical designs of 1945.

UNDERSTANDING SMPTE CODE

Technological advances, unfortunately, can backfire. Before the advent of 1-inch VTRs, which unlike quad machines can display a picture in a shuttle mode, clients had to make detailed notes in order to locate shots on their tapes. Now, a production assistant (PA), confused by eight-digit numbers and knowing a shot can be found "on the fly" in "post," often carries thick binders of "notes" which do not include shot locations or even the name of the reel. What that assistant forgets is that searching in post for shots, even in the shuttle mode, is time-consuming, frustrating and very expensive. High technology does not relieve us of our responsibilities as production professionals. Used properly, technology is a way to accomplish our objectives with the best possible results. Time code is probably the single most misunderstood and misused tool in video production; a sad fact indeed, considering the incredible power of code to streamline and simplify almost every production task.

The first step in the right direction is to realize that SMPTE code was devised as a tool for use by automated machines. In other words, systems are supposed to be reading the code, not people. No director would expect a PA to keep a full and accurate log of scenes, takes and continuity with eight-digit numbers. There is nothing so ridiculous as a person sitting in front of a computer, scribbling away madly with a pencil and paper!

The effective use of automation means *less* scribbling, which results in more detailed notes. Remember the important property of numbers as symbols—they can represent more than a count. Time code can be used to describe much more about footage than merely a frame address.

For example, code can tell you which performance you are viewing. Videotaping *All In The Family* at MetroTapeWest in Hollywood, CA involved shooting the episode in front of a live audience, breaking for dinner, changing the script based on audience feedback, and then shooting the entire show again in front of a different audience. These were called the "dress" and "air" shows. In addition, pickups were shot after the air show to correct fluffed lines or provide more coverage in editing.

The master tapes and offline cassettes were coded with a time of day drop-frame code that was visible in the control room. The dress show, which usually began at about 6:00 p.m., always had code with 18-hour digits. The air show normally started at about 8:00 p.m. and showed 20-hour code addresses. Pickups would begin after the audience was cleared, around 10:30 p.m., with 22-hour code.

By merely interpreting the time of day, a wealth of information was available in the edit decision list (EDL): the amount of rewrite (air/dress ratio), if the shows started on time, the amount of overtime, how much time was spent re-doing scenes in pickups, even the length of the dinner break.

USER BITS—CUSTOMIZING TIME CODE

In taping the show, nothing special was done to manage time code. The code generator was set once, like a clock, and forgotten until a number was useful. This simple interpretation of code addresses as events corresponding to a time of day allowed the code to provide information external to tape frame locations.

SMPTE code can be used to capture and recover information about more complex processes with just as much transparency. To understand time code's usefulness, we should reexamine its basic structure.

Time Code Structure

SMPTE defines time code as a digital signal which is recorded as a series of rectangular pulses. Each pulse represents a bit; 80 bits are recorded for each frame. Twelve bits are assigned as the sync word, which is phase-aligned with tape frame pulse and vertical sync. In this way, code numbers change exactly when the videotape frames change. There is one guard bit on either side of the sync word to isolate it from other data. Two more bits indicate the front of the frame word. These direction-sensing bits tell the computer which way to search for higher or lower addresses. Thirty-two bits are assigned to time of day to indicate hours, minutes, seconds and frames. One of these bits is the drop-frame flag, which indicates drop-frame code. The allocation of this data is strictly controlled by SMPTE for standard interchangeability and may not be changed in any way by the user. This is the time of day (TOD) portion of the code data.

After the TOD portion has been assigned, 32 bits are left available. Rather than assigning these, SMPTE allows users to define them for their own applications. The user data is interleaved in the "spaces beside" the TOD data in a manner similar to the interleaving of chroma and luminance in broadcast transmissions. This arrangement of the "parallel user data" allows two separate addresses to reside within one code word (see Figure 9.2). Code-reading equipment locks on to the non-address data and can decode either TOD or user data as the address.

Unfortunately, not much use has been made of parallel user data. The next example will show just how much can be done with user-equipped code.

MUSIC VIDEO AND CODE

Most music videos, whether shot on film or tape, are posted on video. It is also common practice to lip-sync performers to the finished mix of a track. An audio dub of the mix is played for the artists during each take. If the budget allows and the director is clever, a "scratch" recording of the audio playback is done with each take. Very often, this does not happen and the editor must become a lip-reader to sync these takes.

The importance of the audiotape used for the performers' reference cannot be

Figure 9.2: User (parallel) Data Time Code

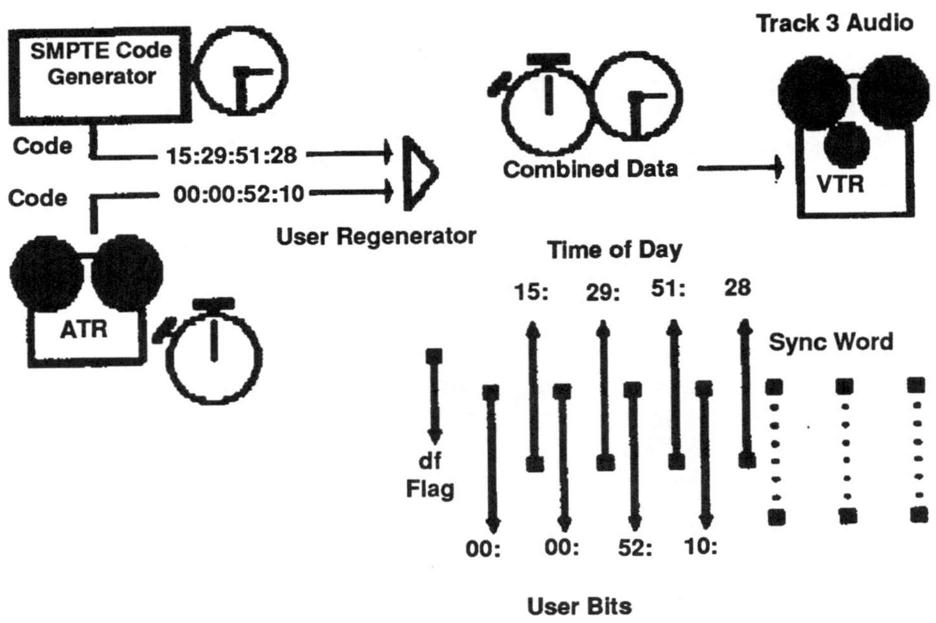

overemphasized. Since the artists' tempo is derived from this tape, it is the "master metronome" for the entire song. If the producers intend to relay the audio mix after editing, the performance reference must be an exact duplicate of that master mix and must play at precisely the same speed. The performance tape should be resolvable to the master, and that is guaranteed by including a synchronizing track—pilotone (60 Hz) or SMPTE code. The location ATR is usually a Nagra or similar resolving recorder that locks to the reference track.

In our example, we will assume that because code is available on film, code rather than piltone has been used. Whether originally shot on tape or film, the same advantages apply.

We will also assume that a film-style single-camera method (the prevalent method for music videos) was used. The director shot multiple takes and moved the camera around for different coverage. TOD code was combined with the code coming from the performance audiotape through a user bit regenerator, or data adder. The ATR's code was inserted as user bits, and the composite code was recorded on track-3 audio of the 1-inch VTR (refer to Figure 9.2).

The TOD count advanced normally as the day wore on, but the user bits were the numbers coming from the audiotape. These were always the same at the same part of the song, regardless of the time of day. In this case, the parallel data indicated the position of the song playback at each instant, which made it almost impossible to lose sync. Even if the audio tracks on the VTR were empty, lip-reading was unnecessary. The user bits served as a "spare" song track.

Though very neat, this use of parallel data has disadvantages. Remember, only one address (TOD or user) can be read at a time. If the video was assembled with parallel data, each edit in the list would be logged as a match-frame to the previous cut. While extremely fast in execution (song sync is automatic), the list would show no distinctions of videotape locations—user bits are the ATR locations. So, auto-assembly is precluded, because no take location is included. The computer cannot tell one take from another. Take identification information is part of the TOD address. This demands manual intervention to locate takes and the writing in big log books of those horrible eight-digit numbers we are trying to avoid.

This dilemma became obvious to the author during the editing of a music video for a group called ToTo. In the following example, he solved the problem while working with Kim Friedman, a director known for his accurate coverage of musical performances.

> "Kim had shot the band with two cameras, each isolated into its own VTR. Because he made certain that every cymbal tap and drum hit were photographed, Kim was half submerged in a sea of notes—very good ones—locating each shot with code.
>
> Although the edit was going quite well, Kim heaved a big sigh and muttered, "I wish these damn numbers were smaller, at least I could find my cup of coffee!" The picture of my friend losing his coffee mug amidst an ocean of paper haunted me that night and I started thinking like a lifeguard. An effective "life preserver" would have to do more than user bits. Kim had expressed the exact nature of the problem: the numbers were too big. If I could imagine a way to make them smaller, his notes would shrink to a less threatening size.
>
> Since numbers can represent anything, we could change the meaning of the user "hours" to "takes" by intercepting the two most significant digits during recording, and manually advancing them via thumbwheels each time we rolled the ATR for the next shot (see Figure 9.3). Though the six other digits were still important for maintaining song synchronization, they could be ignored in the shot log. The machines would use all eight digits; the director need only concern himself with the first two—the take numbers. Those were the only numbers written in his shot book.
>
> Now the TOD data could be ignored. The computer could find, list and execute edits by user address under auto-assembly. The song sync was intact, and the PA could record one or two digit take numbers in a notebook that was suddenly slim and easy to read. When Kim needed a shot, he merely asked for take 11, not 11:14:36:22 (see Figure 9.4).
>
> The last six numbers of the address are used exclusively by the machines for keeping song sync; the first two numbers are the only ones the humans

Continued on page 131

Figure 9.3: Standard User Bits and SuperBits

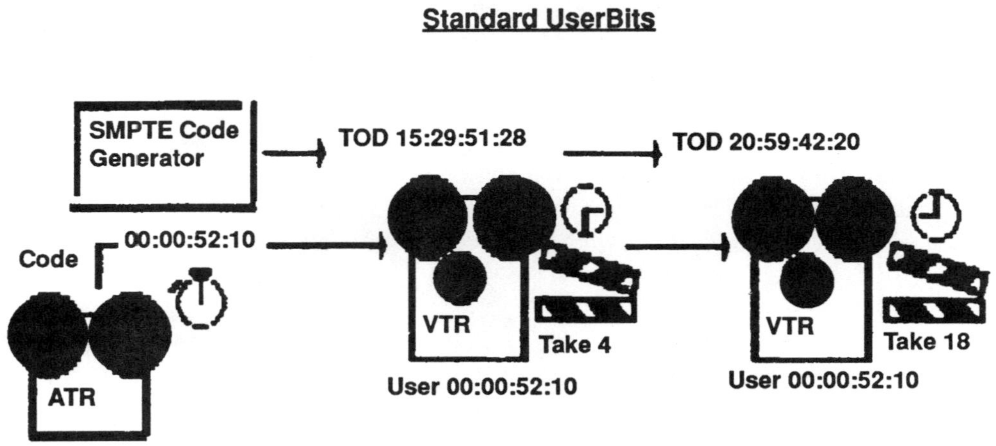

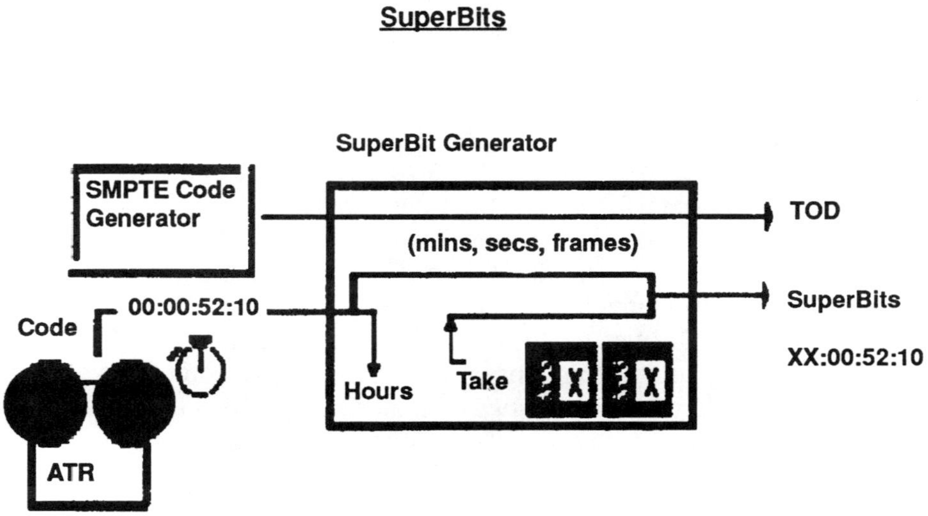

Figure 9.4: Edit Decision Lists/Code Formats

<u>TIME OF DAY</u>
FCM: Drop Frame

001	001	V	C	14:24:55:10	14:24:55:22	01:00:00:00	01:00:00:12
002	001	V	C	11:24:22:22	11:24:24:28	01:00:00:12	01:00:02:18
003	001	V	C	13:00:01:28	13:00:02:28	01:00:02:18	01:00:03:18
004	001	V	C	09:01:25:06	09:01:35:06	01:00:03:18	01:00:13:18

<u>USER BITS</u>
FCM: Drop Frame

001	001	V	C	00:00:52:10	00:00:52:22	01:00:00:00	01:00:00:12
002	001	V	C	00:00:52:22	00:00:54:28	01:00:00:12	01:00:02:18
003	001	V	C	00:00:54:28	00:00:55:28	01:00:02:18	01:00:03:18
004	001	V	C	00:00:55:28	00:01:05:28	01:00:03:18	01:00:13:18

<u>SUPER BITS</u>
FCM: Drop Frame

001	001	V	C	18:00:52:10	18:00:52:22	01:00:00:00	01:00:00:12
002	001	V	C	07:00:52:22	07:00:54:28	01:00:00:12	01:00:02:18
003	001	V	C	11:00:54:28	11:00:55:28	01:00:02:18	01:00:03:18
004	001	V	C	05:00:55:28	05:01:05:28	01:00:03:18	01:00:13:18

need to worry about.

That's automation. I can get the tapes cued faster with less typing, and Kim never loses his coffee."

CONCLUSION

The examples discussed in this chapter show how many little problems can crop up in a complex system like television. There were many loose ends left dangling in the standards adopted by the NTSC back in the 1940s and 1950s. While we are dealing with a few of the technical difficulties, many problems still remain. They persist because of the usefulness of the broadcast system and our need to keep our television signals available to everyone.

The system will only become more complicated as it evolves. As more subsystems are interfaced to television, we see more people specializing. There are, for instance, many graphic artists working on digital painting systems who have little or no idea of how their output affects others working on the same production. They may not know how to use time code or even how to transfer their signals properly to videotape.

While no one can know all there is to know about a system as big as NTSC television, it is vital that everyone who attempts to work within the system know its basic structure.

The business we are in is, first and foremost, *communication*. We have as our objective, the useful transmission of messages to large numbers of people. That fact will not change. Before we can hope to communicate with all the viewers, we must be able to communicate with each other. Our knowledge of our medium provides the language that serves that purpose.

That knowledge starts with the fundamental material we have covered in this book. The most exciting aspect of working in television is realizing that the learning process goes on.

Glossary

A mode: Linear method of assembling edited footage; the editing system performs edits in a numerical sequence, stopping whenever the edit decision list calls for a reel that is not assigned to a VTR.

amplitude: Height of a signal (in volts).

analog recording: The common form of magnetic recording in which the recorded waveform signal maintains the shape of the original waveform signal.

aspect ratio: Height to width of the displayed picture; in video, the standard aspect ratio is 3:4.

assemble edit: Electronic edit that replaces all previously recorded material with new audio and video and a new control track, starting at the edit-in point.

back porch: Portion of a composite video signal between the trailing edge of horizontal sync pulse and the leading edge of the video portion of the signal.

bit: Short for binary digit; the smallest unit of digital information with a value of 0 or 1.

B mode: "Checkerboard" or non-sequential method of auto-assembly in which the computerized editing system performs all edits from the reels that are currently assigned to VTRs, leaving gaps that will be filled by material from subsequent reels.

breezeway: Portion of the video signal's back porch between the trailing edge of horizontal sync and the start of color reference burst.

byte: A group of adjacent bits, which combine to form one unit of data.

capstan: Rotating shaft on a videotape recorder transport path. The capstan controls the tape speed as it travels from the supply reel to the tape-up reel.

cathode ray tube (CRT): TV picture tube.

chroma key: Method of removing a specific color from an image, usually the blue or green, to cut a hole for the new image.

chroma: Video color.

chrominance: The portion of the TV signal that contains color information.

clamping: A video processing operation that provides a line-by-line correction of the video blanking or sync tip level to a fixed DC reference voltage.

clipping: Imposing an electronic limit to avoid overdriving the audio or video portion of the television signal.

C mode: Non-sequential method of assembly in which the edit decision list is arranged by source reel number and ascending source time code.

color bar test signal: Standard color test signal, displayed as rows ("bars") of color.

color black: Video signal containing horizontal sync, color reference burst and pedestal.

color reference burst: Color synchronizing signal inserted on the back porch of horizontal sync. When compared with the color subcarrier signal, the color reference burst determines the hue of the video image.

color subcarrier: Carrier frequency (3.58 MHz in NTSC, 4.43 MHz in PAL) on which the color signal information is impressed.

colorframe: Condition at edit point where the color reference burst phase of the incoming shot is very close to or the same as the phase of the outgoing shot.

component video: Video signal in which the luminance and sync information are recorded separately from the color information.

composite video: Video signal containing both picture and sync information.

control track: Portion of the video recording used to control the longitudinal motion of the tape during playback.

cut: To instantly replace one scene with another. Also called a take.

demodulation: To strip a signal of the carrier frequency onto which it was modulated; a signal is demodulated after it has been broadcast and prior to its display.

digital: Electronic system which functions by converting the analog signal into a series of discrete binary bits.

digital framestore: Memory device that scans and stores a complete frame of video information after the information has been converted from analog to digital form. Digital framestores are used to create digital special effects.

dissolve: A video transition in which the existing image is partially or totally replaced by superimposing another image.

drop-frame code: SMPTE time code format that skips two frames per minute except on the tenth minute, so the time code stays coincident with real time.

dub: To make a copy.

edit decision list (EDL): List of edits performed during offline editing.

electromagnetic radiations: The broad spectrum of radiations between radio waves and cosmic rays.

encoding: Adding technical data, such as time code, to a video recording.

fade: Dissolve to or from black.

fader: A sliding potentiometer (pot) control that diminishes the audio or video signal.

field: Half of one television frame. In NTSC 262.5 horizontal lines at 59.94 Hz.

film chain: A total system comprised of a film projector, slide projector and video camera, used to convert film or slides to video. Also called telecine.

flying erase head: Moving erase heads mounted near or next to the video heads on a videotape recorder. Flying erase heads are activated during an insert edit.

frame: One complete video picture. A frame contains two fields, scanned at the NTSC rate of 30 frames per second or the PAL rate of 25 frames per second.

function generator: Oscillator capable of outputting a wide variety of differently shaped waveforms all phase-locked to sync.

genlock: Device synchronizing video signal sources.

harmonics: Frequencies that are exact multiples of the original frequency.

head: Magnetic pickup device used in a videotape recorder to record, erase or reproduce video and audio signals.

helical VTR: Type of VTR in which the videotape is wrapped partially or completely around the video head assembly. The video signal is recorded across the tape in slanted strips. Also called slant-track VTR.

horizontal sync: The portion of the composite video signal that synchronizes the scanning electron beam of the TV monitor so that each line of picture information will start at the same lateral position during scanning.

hue: The shade of a color.

incident color: Color that is emitted directly from a source.

insert edit: Electronic edit in which a new segment is inserted into program material already recorded on the videotape; the control track is not replaced during the editing process.

interlace scanning: NTSC television scanning process in which two fields of video are interlaced to create one full frame of video.

key: Switcher transition in which graphics are inserted over a scene or in which one video image is placed into another.

kinescope: A method for making a film copy of a TV program by pointing a movie camera at a television screen displaying the program.

luminance: Amplitude (strength) of the gray scale portion of the television signal.

matte: Form of key in which text appearing as color-filled letters is keyed by a luminance signal, usually from a monochrome camera.

mix/effects bus: Grouping of modules that form a complete processing unit of switching components. Several buses may be cascaded together to create complex, multiple layers of effects.

modulation: Impressing a signal onto a carrier frequency for transmission purposes.

monochrome: Black and white.

non-composite video: Video signal that does not contain horizontal and vertical sync pulses.

non-drop frame code: SMPTE time code format that continuously counts a full 30 frames per second. As a result, non-drop frame time code is not coincident with real time.

NTSC: National Television System Committee, the group that established the color TV transmission system in the United States.

NTSC color video standard: The U.S. standard for color TV transmission calling for 525 lines of information, scanned at a rate of 30 frames per second.

NTSC primary colors: Colors that cannot be duplicated by the combination of any other colors (red, green and blue).

NTSC secondary colors: Those colors which result from the addition of two primary colors: magenta (red + blue), yellow (red + green) and cyan (green + blue).

offline editing: Preliminary post-production session, used to establish editing points and to prepare an edit decision list.

online editing: Final editing session. The stage of post-production in which the edited master is assembled from the original production footage, usually under the direction of an edit decision list.

PAL: Phase Alternating Line, a color TV standard used in many countries. PAL consists of 625 lines scanned at a rate of 25 frames per second.

pedestal: Reference black level of the video signal (7.5 units for color, 10 units for black and white). Also called setup.

phase: The relative timing of a signal in relation to another signal; if both signals occur at the same instant, they are said to be in phase.

piezoelectric crystal: A salt crystal capable of delivering an electrical signal output when it is twisted or flexed. Video heads can be mounted on piezoelectric crystals and mounted into head cylinders. If a voltage is applied to the crystal it will flex, thereby changing the physical position of the head. This allows the head to properly trace the video tracks when the videotape is played back.

pinch roller: A rubber roller on the tape transport path that presses the videotape against the capstan.

pixel: Picture element, the basic building block of the video image.

prelay: The phase of audio post-production during which music, sound effects, dialogue replacement and announce tracks are added to the master multitrack before the final mix.

pulse: Variation of a constant signal for a certain period of time.

quadruplex (Quad): Videotape system that records with four video heads mounted 90° apart around the rim of a rotating head wheel. Quad VTRs usually use 2-inch videotape. Also called transverse scan VTR.

quard band: Blank section of tape used to separate tape tracks.

radio frequency (RF): High frequency portion of the electromagnetic spectrum used for transmitting television and radio signals.

raster: Area of the TV picture tube that is scanned by the electron beam. Also, the visual display present on a TV picture screen.

real time: Actual clock time in which events occur.

reflected color: Color that bounces off an object.

resolution: Amount of detail in the video image, measured along both the horizontal and vertical axes.

RGB: Red, green and blue, the primary colors in color video.

saturation: The amount of color in the television picture.

scanner assembly: Mechanical drum assembly that contains the rotating heads and tape guiding elements used to record and play back helical VTR recordings.

SECAM: Sequential Colour a Memoire, a TV standard developed by the French and used in France, the USSR and other countries.

self key: Key effect in which text appears superimposed over the picture.

servo: Error detection circuit in the VTR responsible for maintaining proper tape speed.

sweetening process: Process of mixing sound effects, music and announcer audio tracks with the edited master tape's audio track. Also called audio post-production for video.

switcher: Unit that allows the operator to switch between video camera signals; is responsible for special effects and controlling what images are sent to and from various sources.

sync generator: Component that generates sync pulses.

sync: Synchronization—usually refers to the pulses necessary to coordinate the operation of several interconnected video components.

time base corrector (TBC): Electronic device used to correct video signal instability during the playback of videotaped material.

time base error: Variation in the stable relation of picture information, color information and video sync pulse during the VTR playback process.

time code: Electronic indexing method used for editing and timing video programs. Time code denotes hours, minutes, seconds and frames elapsed on a videotape.

vertical sync: Sync pulses that control the vertical field-by-field scanning of the front surface of the video camera's pickup tube by the electron beam.

videotape: Oxide-coated, plastic-based magnetic tape used for recording video and audio signals.

wavelength: The length (in meters) of one cycle of a signal.

wipe: Special effect transition in which a margin or border moves across the screen, wiping out the image of one scene and replacing it with another.

writing speed: Writing speed in video recorders is determined by how fast the video heads move across the tape, in addition to the tape speed.

Index

Abekas, 93, 115
Active signals, 25
Alpha head wrap, 90
Ampex, 56, 64, 69, 71-72, 73, 76, 78, 80, 83, 86-87, 91, 92, 96, 112, 114, 115
Ampex Time Element Compensator (AMTEC), 72
Arri Corp., 102
Assemble editing, 78-79
AT&T, 50
Audio for film, 101-103
Audio for video, 96-97
  layover process, 96
  layback, 97
  mix, 97
  prelay, 97
Audio sweetening, 95-104

Bell Laboratories, 108
Beta format, 73
  Betacam, 91
  Betamax, 93, 100, 101
  Beta Hi-Fi, 100
Bing Crosby Productions, 63
Black burst, 42
Bosch, 90, 114

Callaway Engineering (CE), 118
Cathode ray tube (CRT), 4-6, 9-10, 11
CBS, 83
Chrominance, 2, 21, 29-33, 123
CMX Systems, 83-86, 93, 118
Coherent Communications, 102
Color bar test signal, 42
Color basics, 28
  incident colors, 28
  NTSC primary and secondary colors,
  reflected colors, 28
Color burst, 33-34
Color camera, 29-30
Colorframing, 121-123
Color time base, 36-37
Color time element compensato
ORTEC), 72
Composite video, 49
Control track, 69-70
  capstan tach, 70
  control track signal, 70
  headwheel tach, 69
  home track, 69
  reference head, 69
Convergence Corp., 93, 114

Decca Co., 63
Desilu Studios, 80
Digital Equipment Corp. (DEC), 84
Digital framestore, 50-51
Digital recording, 92-93, 105-119

Digital videotape recorders (DVTRs), 114-115
Direct recording process, 57-59
  magnetic tape, 57-58
  microphone, 57
  recording head, 57
Double-system sync sound, 95-96
DroidWorks, 94, 116

Eastman Kodak, 102-103
Edit decision list (EDL), 82-83, 86-87, 93, 126, 131
Editec, 79-80, 96
Edivue, 76
8mm format, 73
Electricity, 2-3
  amplitude, 3
  applying signals, 2-3
  current, 2
  electrical signals, 2-3
  resistance, 2
  voltage, 2
  wavelength, 3
Electromagnetism, 56
  electrical field, 56
  magnetic field, 56
Electronic Editing Corp. (EECO), 80, 94
E-Mu Systems, 100, 116
Eniac machine, 106-107
Eventide 949 harmonizer, 110-112

Fairlight Corp., 100, 116
Federal Communications Commission (FCC), 21, 30
Flying erase head, 78, 89
FM recording, 65

Genlock, 50
Grass Valley Group (GVG), 86, 114
Guard band, 69

Helical format, 72-73
  helical scan, 73, 76
High-definition television (HDTV), 14, 104
Horizontal shift, 121
  troubleshooting of, 121-124

Insert edit, 79-80
Integrated switching, 53
Integrated Systems Corp. (ISC), 93
Interlaced scanning, 12-14, 19

JVC, 72

Keys, 43
  chroma key, 45, 47
  hold-out matte, 45
  mask key, 47
  mattes, 45
  self key, 43
  using two or more keyers, 45
Kinescope, 9-10, 15, 16, 28, 31, 55, 75

Light, 1-2
  chrominance, 2
  electromagnetic radiations, 1
  hue, 2
  luminance, 2
  photons, 1
  saturation, 2
Linear scanning, 14-21
  creating the scanning pulse, 14-17
  photoemissive electron, 15
LucasFilms, 93, 94, 116
Lucy leader, 80
Luminance, 2, 10, 14, 16, 21, 29-35, 6

MII Recam, 91
Magnetic video recordings, 62-63
  direct longitudinal, 62
  longitudinal VTRs, 63
  stationary heads, 62
Mark I electronic editing accessory, 78
Mark II electronic editor, 79
Matrix switcher, 51-53
  Central Dynamics, 51, 53
  Grass Valley Group, 51
Memorex, 83, 84
M.I.T. Media Lab, 105
Mix/effects bus, 42, 47
Monochrome transmission characterist
  FCC, 30

Moog synthesizer, 98
Music videos, 127-131

Nagra audio recorder, 91, 95
National Television System Committee (NTSC), 11-14, 19, 23, 25-37, 71, 76, 81, 91, 92, 118, 121
NBC, 9
New England Digital, 100, 116
Non-composite video, 49
Non-linearity in media, 59-60
  saturation point, 60
NTSC color, 25-37
  encoded color, 30-36
Nyquist Criterion, 110

Odd-line interlace, 12-14
Offline editing, 86-87, 93-94
Omega head wrap, 90-91
Orrox Co., 84

PAL, 11, 12
Panasonic, 86
Panavision Corp., 103
Phase modulation (PM), 32-33
  color burst, 33
  color sync, 33, 36
  polar mapping, 32
  vector analysis, 32
Piezoelectric crystal, 89-90
Pilotone sync, 95-96, 102
Pinch roller, 70, 73
Pixel, 11-12, 16, 17

Quad recording cycle, 65-69
  servo, 66-67
  track pattern, 67
  time base, 67-69
Quadruplex machine, 64-65
  head assembly, 64
  headwheel, 64
  transverse scan, 64, 76
Quantel, 114, 115

Radio frequency (RF), 25-28, 30
  amplitude modulation (AM), 26-27
  demodulation, 26
  frequency modulation (FM), 27
  NTSC regulation, 27-28
  RF envelope, 26
  transmission of, 25-26, 30
Raster, 11, 19, 43
RCA, 9, 72, 73
Recording before tape, 55-56
  radio, 55
  television, 55
Record gap limitations, 60-62
  equalization, 62
  post-emphasis, 62
  pre-emphasis, 62
Resolution, 11, 14
The Roland Co., 99

Scanning CRT, 9-10
  luminance, 10
  video signal, 10
SECAM, 11
Signal processing, 49-50
  composite video, 49
  non-composite video, 49
Signal timing, 39-41
  cable length, 40-41
  master clock pulse, 40
  pulse distribution system, 40
  sync generator, 40
SMPTE, 21, 76, 81, 82, 91, 92, 95, 100, 1
SMPTE Standard RS-170A, 121, 124
SMPTE type-C 1-inch format, 87-90
  auto scan, 89
  benefits of, 90
  evolution of, 91-93
  helical scan, 87-89
  scanner, 87
Sony Corp., 72, 86, 87, 91, 100, 115
Sound, 1
  loudness, 1
  pitch, 1
  pressure waves, 1
Soundstream Corp., 100
Subcarrier-to-horizontal sync, 123-124

Switching, 41-50
crosspoint, 41-42
mix/effects bus, 42, 47
summing or mixing, 41-42
Switcher console, 45-46
Switcher transitions, 42-45
cuts, 42-43
dissolves, 43
wipes, 43
Sync, 17-21
horizontal, 17-18
vertical, 19-21
Synchronizing signals, 25

Time base stability, 71-72
color subcarrier, 72
time base corrector, 72
Time code, 81-83, 122
customizing of, 126-127
drop-frame, 81
longitudinal, 83
non-drop frame, 81
structure of, 127
vertical interval, 83
Timing and phasing, 42
black burst, 42
color bar test signal, 42
Triple reentry downstream switcher, 46
crosspoint, 47
fader, 47
potentiometers, 47

U-matic format, 73, 83

Vertical blanking, 19-22, 39, 76
VHS format, 73, 101
Video production recorders (VPRs), 73
Videotape editing, 75-94
automation of, 80-81
cutting and splicing, 76
edivue, 76
Vital Industries, 114

Waveform, 3-6
oscillograph, 3-4
Writing speed, 59

About the Author

Michael A. Krupnick has been involved in broadcasting since 1964, when he worked after school as a disc jockey, reporter and engineer at two different radio stations in Columbus, OH.

Majoring in theatre in Miami, FL, he joined WTVJ television as an audio engineer and eventually became one of the top-rated videotape editors in that market. He worked on such shows as CBS's *Sixty Minutes*, the NASA Apollo space launch coverage and network Superbowl broadcasts.

In 1974, Mr. Krupnick went to San Francisco where he studied holography, produced two radio plays for Bay Area National Public Radio, and worked with Firesign Theatre.

In 1976, he joined the ABC network videotape department in Hollywood. He later was hired by MetroTapeWest, where he worked for four years on such shows as *All In The Family, The Jeffersons, Three's Company, One Day At A Time* and *Diff'rent Strokes.* At this time he also became interested in music videos.

In 1981, Mr. Krupnick became a freelance editor/mixer, working on a wide variety of projects at many different facilities in Hollywood. During that same year, he was approached by the University of Southern California to teach a class for the graduate students of the film department on television production. He designed and submitted a curriculum on technical theory. This course emphasized facets of broadcasting and production previously uncovered in other classes at the university. It was greeted with enthusiasm, and was solicited by the Los Angeles Institute of Video Technology and MetroMedia, Hollywood. Mr. Krupnick began teaching the class part-time as a 13-week course at those schools and quest-lectured at UCLA. The class eventually ran for seven years at MetroMedia, where it was offered tuition-free as a public service to students who could not afford to enroll at other schools. Mr. Krupnick also served as a member of the Editors' Advisory Board assembled by the CMX Corporation to help design the 3400 editing system. He has also served as instructor for CMX operations classes for clients and users.

Today, Mr. Krupnick owns and operates The Mixmasters Studios in Hollywood, which services a wide variety of clients with audio and video post-production for commercials, videos and feature films.